Multimedia-enabled Sensors in IoT

Data Delivery and Traffic Modelling

Multimedia-enabled Sensors in IoT

Data Delivery and Traffic Modelling

Fadi Al-Turjman
Antalya Bilim University, Turkey

CRC Press
Taylor & Francis Group
Boca Raton London New York

CRC Press is an imprint of the
Taylor & Francis Group, an **informa** business

MATLAB® is a trademark of The MathWorks, Inc. and is used with permission. The MathWorks does not warrant the accuracy of the text or exercises in this book. This book's use or discussion of MATLAB® software or related products does not constitute endorsement or sponsorship by The MathWorks of a particular pedagogical approach or particular use of the MATLAB® software.

CRC Press
Taylor & Francis Group
6000 Broken Sound Parkway NW, Suite 300
Boca Raton, FL 33487-2742

First issued in paperback 2020

© 2018 by Taylor & Francis Group, LLC
CRC Press is an imprint of Taylor & Francis Group, an Informa business

No claim to original U.S. Government works

ISBN-13: 978-0-367-57174-0 (pbk)
ISBN-13: 978-0-8153-8711-4 (hbk)

Library of Congress Cataloging–in–Publication Data

Names: Al-Turjman, Fadi, author.
Title: Multimedia-enabled sensors in IoT data delivery and traffic modelling / Fadi Al-Turjman.
Description: Boca Raton, FL : CRC Press, Taylor & Francis Group, 2017. | Includes bibliographical references.
Identifiers: LCCN 2017048014 | ISBN 9780815387114 (hb : acid-free paper)
Subjects: LCSH: Internet of things. | Wireless sensor networks. | Telecommunication--Traffic.
Classification: LCC TK5105.8857 .A43 2017 | DDC 006.2/5--dc23
LC record available at https://lccn.loc.gov/2017048014

Visit the Taylor & Francis Web site at
http://www.taylorandfrancis.com

and the CRC Press Web site at
http://www.crcpress.com

To whom I owe the true love,

To whom the depths of my soul are thundering, and the poem is created…

To the most important woman in my history, who thinks the same without speech

and speaks the same without thinking…

To my wonderful wife,
Fadi Al-Turjman

Contents

Author

Dr. Fadi Al-Turjman is a professor at Antalya Bilim University, Antalya, Turkey. He earned his PhD in computing science from Queen's University, Canada, in 2011. He is a leading authority in the areas of smart/cognitive, wireless and mobile networks' architectures, protocols, deployments, and performance evaluation. His record spans more than 150 publications in journals, conferences, patents, books, and book chapters, in addition to numerous keynotes and plenary talks at flagship venues. He has received several recognitions and best paper awards at top international conferences and led a number of international symposia and workshops in flagship ComSoc conferences. He is serving as the lead guest editor in several journals including the *IET Wireless Sensor Systems* (WSS), *MDPI Sensors* and Wiley. He also served as the general workshops chair for the IEEE International Conference on Local Computer Networks (LCN'17). Recently, he published *Cognitive Sensors & IoT: Architecture, Deployment, and Data Delivery* with Taylor & Francis, CRC, Boca Raton, Florida. Since 2007, he has been working on international wireless sensor networks (WSNs) projects related to remote monitoring, as well as Smart Cities related deployments and data-delivery protocols using integrated RFID (Radio Frequency Identification)-Sensor Networks (RSNs).

Chapter 1

Introduction

Recent developments in wireless sensor network (WSN) technology (which enables communications ranging from a few meters to city-scale applications in order to perform simple tasks such as sensing, actuation, and computing) have promoted a new class of multimedia applications in multidisciplinary domains, which significantly depend on sensing technologies [1]. The dramatic development of sensing/communication techniques makes it the most appropriate technology in the Internet of things (IoT) paradigm. The recent advancements of WSNs in the IoT have been widely promoted in environmental, industrial, and biomedical sensing and monitoring applications, which significantly depend on real-time data [1]. Consequently, a new technology called wireless multimedia sensor networks (WMSNs) has emerged to achieve more reductions and savings in terms of multimedia gathering cost. WMSNs can gather and deliver multimedia data such as sound streams, images, video, and scalar data (e.g., temperature/humidity readings). WMSNs have not only enhanced the existing applications of traditional WSNs but have also enabled new services. However, there are several restrictions facing their design and implementation in reality while satisfying the desired quality of service (QoS) [2]. In this book, we focus on the constraints related to QoS requirements and what necessitates efficient routing protocols. Such routing protocols are needed for real-time and non–real-time multimedia applications. However, the former requires different QoS metrics and design aspects, which are supposed to be considered at the data networking/routing protocols. Consequently, WMSNs have evolved in the literature with a special kind of routing protocols called "multipath" routing protocols [3]. Multipath routing can be considered as a new technique in delivering data via several routing paths. Unlike the single-path routing, multipath routing protocols provide "more" adequate network resources, improve the packet arrival ratio of multimedia contents, and guarantee the QoS at high data rates [3]. This kind of protocol is cost-effective for overloaded networks and lasts for prolonged lifetime periods. Accordingly, in this book, we investigate the importance of multipath routing strategies that assure the QoS in the IoT paradigm while contrasting their pros and cons, mechanisms for multipath routing, classifications in terms of the multimedia delivery exploitation, and design issues.

Most of the literature surveys about WMSNs routing protocols care about the multimedia application and the network architecture requirements while ignoring significant criteria, such as the lifetime of the WMSN and key QoS requirements in these networks. The authors in Reference 4 discussed constraints and solutions related to each layer of the WSN stack. In Reference 5, the authors reviewed several routing protocols and a classification based on multipath availability, query type, and in-network negotiation was proposed. In Reference 6, a taxonomy was developed and certain design issues such as energy efficiency, geographic location, network scalability, and cross-layer approaches were considered. The authors in Reference 2 introduced protocols that support real-time applications in WSNs. Existing routing strategies were classified based on the network structure, communication model, and topology shape. Position-based routing algorithms and metrics like energy consumption, negotiation overheads, complexity, reliability, scalability, and multipath strategies were considered in Reference 7. In Reference 8, scalability issues and details of various protocols, merits, and demerits were presented. Constraints and factors that influence the multimedia delivery, cross-layer design optimization, and performance evaluation were investigated in real-time multimedia WSNs.

In fact, the majority of research attempts in WMSNs are made to achieve energy-efficient multipath selection. Energy-efficient routing strategies were introduced in Reference 9, whereas the main characteristics of choosing the most appropriate multipath routing schemes and their classifications were discussed in Reference 10. Alternative path routing, reliable data transmission, and efficient resource utilization were the three major criteria in which multipath routing protocols were classified in Reference 11, whereas a different classification based on selection techniques and traffic distribution mechanisms was presented in Reference 12. In Reference 13, the Internet, end users, and Internet service providers (ISPs) were used in multipath routing and provisioning.

This book covers all the aforementioned types of routing protocols while classifying them based on the multipath selection criteria. It aims at providing a comprehensive framework spanning the multimedia delivery aspects, traffic modeling, and real case scenarios in the IoT era. In this framework, the end user must be able to exchange multimedia traffic based on the required QoS attributes. The WMSN itself must be able to dynamically adapt for the varying network/user conditions and provide fault-tolerant multipath routes for the handled multimedia traffic as indicated in Reference 14.

1.1 Contributions

We propose the multipath routing approach as a key solution for multimedia delivery in large-scale IoT applications. We assume a WMSN within which super and light nodes are introduced. These super nodes are able to make delivery decisions that dynamically adapt to user requirements and network conditions. Toward this end, our main contributions in this book can be summarized as follows:

1. We start with a comprehensive overview for the multipath routing techniques and related works in the literature. We identify critical QoS assurance factors and classify various techniques that can be applied to sensor networks in IoT applications.
2. We combine the use of light cost-effective WSNs and the smart paradigms in the IoT, which can potentially support mobile multimedia applications. This combination supports and resolves key IoT design aspects related to mobility and multimedia traffic in a cost-effective manner.
3. According to the proposed WMSN architecture, we investigate the data routing problem while stressing the fault-tolerance design factor. We propose multipath routing strategies that guarantee the QoS in multimedia delivery.
4. We propose two heuristic data delivery approaches that could be the most appropriate for multimedia delivery in this work to cope with the next generation of IoT trends. The proposed data delivery approaches either help to choose paths that deliver data with the least delay toward the sink or identify data delivery paths that are more energy-aware and cost-efficient. Both approaches have been discussed in detail and have shown promising results.
5. We analyze and quantify the network traffic of the aforementioned WMSN under realistic operational conditions. Solid mathematical models are recommended and endorsed based on analytical studies.
6. Moreover, we investigate the most appropriate traffic modeling techniques in the IoT and future Internet era. We characterize data requests based on content demand ellipse (CDE), focusing on efficient content access and distribution as opposed to mere communication between data consumers and publishers.

1.1.1 Book Outline

The rest of this book is organized as follows. In Chapter 2, we delve into an overview for the field of multipath routing in the IoT era. In Chapter 3, we provide the details of our optimized multipath routing approach for WMSNs in the IoT paradigm. Chapter 4 provides an energy-aware data delivery framework for safety-inspired multimedia in mobile IoT. Chapter 5 introduces a delay-tolerant framework for integrated radio-frequency identification (RFID) and sensor networks (RSNs) in the IoT era. Chapter 6 introduces a use case in which unmanned air vehicle (UAV)-enabled WSNs for multimedia delivery in safety-inspired mobile IoT is proposed and discussed. In Chapter 7, we assess a common protocol in WSNs, called the duty-cycled asynchronous X-MAC protocol, for dynamic sensor networks such as the vehicular network. In Chapter 8, we provide a mobile traffic model for WMSNs in the IoT. In this chapter, we provide a dynamic model for IoT-specific paradigms to improve the end user satisfaction. Chapter 9 debates a generic information-centric framework for traffic modeling and optimization in the IoT. Finally, we conclude this book in Chapter 10 with potential perspectives on the proposed work and discuss the directions of future work.

References

1. F. Al-Turjman, Cognition in information-centric sensor networks for IoT applications: An overview, *Annals of Telecommunications Journal*, vol. 72, no. 1, pp. 1–16, 2016.
2. M. Z. Hasan, H. Al-Rizzo, and F. Al-Turjman, A survey on multipath routing protocols for QoS assurances in real-time multimedia wireless sensor networks, *IEEE Communications Surveys and Tutorials*, vol. 19, no. 3, pp. 1424–1456, 2017.
3. F. Al-Turjman, Cognitive routing protocol for disaster-inspired internet of things, *Future Generation Computer Systems*, 2017. doi: 10.1016/j.future.2017.03.014.
4. N. Al-Karaki and A.E. Kamal, Routing techniques in wireless sensor networks: A survey, *IEEE Wireless Communications*, vol. 11, no. 6, pp. 6–28, 2004.
5. A. Boukerche, B. Turgut, N. Aydin, M. Ahmad, L. Bölöni, and D. Turgut, Routing protocols in ad hoc networks: A survey, *Computer Networks*, vol. 55, no. 13, pp. 3032–3080, 2011.
6. M. Korkalainen, M. Sallinen, N. Kärkkäinen, and P. Tukeva, Survey of wireless sensor networks simulation tools for demanding applications, in *Proceedings of the IEEE International Conference on Networking and Services*, Valencia, Spain, pp. 102–106, 2009.
7. Z. Jin, Y. Jian-Ping, Z. Si-Wang, L. Ya-Ping, and L. Guang, A survey on position-based routing algorithms in wireless sensor networks, *Algorithms*, vol. 2, pp. 158–182, 2009.
8. Z. Hamid and F. Hussain, QoS in wireless multimedia sensor networks: A layered and cross-layered approach, *Wireless Personal Communications*, vol. 75, no. 1, pp. 729–757, 2014.
9. S. Ehsan and B. Hamdaoui, A Survey on energy-efficient routing techniques with QoS assurances for wireless multimedia sensor networks, *IEEE Communications Surveys & Tutorials*, vol. 14, pp. 265–278, 2012.
10. M. Masdari and M. Tanabi, Multipath routing protocols in wireless sensor networks: A survey and analysis, *International Journal of Future Generation Communication and Networking*, vol. 6, no. 6, pp. 181–192, 2013.
11. M. Radi, B. Dezfouli, K. Abu Bakar, and M. Lee, Multipath routing in wireless sensor networks: Survey and research challenges, *Sensors*, vol. 12, pp. 650–685, 2012.

12. E. David, N. Ndih, and S. Cherkaoui, On enhancing technology coexistence in the IoT Era: ZigBee and 802.11 case. *IEEE Access*, vol. 4, pp. 1835–1844, 2016.
13. F. Al-Turjman, Cognitive caching for the future fog networking, *Elsevier Pervasive and Mobile Computing*, 2017. doi: 10.1016/j.pmcj.2017.06.004
14. F. Al-Turjman, H. Hassanein, and M. Ibnkahla, Efficient deployment of wireless sensor networks targeting environment monitoring applications, *Computer Communications*, vol. 36 no. 2, pp. 135–148, 2013.

Chapter 2

A Survey on Multipath Routing Protocols for QoS Assurances in Real-Time Wireless Multimedia Sensor Networks

2.1 Introduction

The demand for a wide variety of network services and various multimedia applications has been the major driving force behind the innovation and development of various networking technologies such as IEEE 802.11n, 4G/5G long-term evolution-advanced (LTE-A), and wireless sensor networks (WSNs) [1]. Integration of these applications in modern networking poses a new set of constraints on quality of service (QoS) and often requires suitable routing strategies. Routing strategies are key for meeting the different demands for network capacity provisioning and QoS guarantees in such networks. Wireless multimedia sensor networks (WMSNs) have been used in numerous applications in the era of the Internet of things (IoT) that require ubiquitous access to both real- and non-real-time applications as depicted in Figure 2.1 [2]. WMSNs utilize the multimedia sensing technology to monitor changes in the surrounding environment and route the collected information via routing protocols to remote controlling units [3]. WMSNs are expected to be among the pillars in realizing the IoT paradigm by fostering applications such as smart cities, smart agriculture, and smart

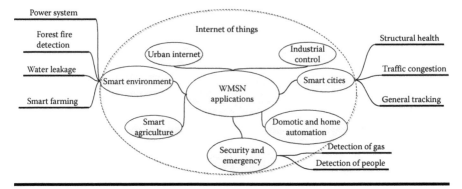

Figure 2.1   Applications of WMSNs in different environments.

security and emergency systems [4]. Certain applications of WMSNs may operate for several months or even years [5]. The deployment of multimedia devices depends on the application and the environment [5]. This means that each sensor node should be modified distinctively according to the application scenarios [6]. For example, in security and emergency applications, low-cost maintenance is required, whereas urban applications are more apt to risk of environmental interference [5]. Therefore, the network designer may encounter several conflicting design aspects, challenges, and factors that influence the design of WMSNs in these environments. These characteristics, challenges, and factors depend on the nature of real-time multimedia traffic data flow, such as the application specifics of QoS requirements, high bandwidth, tolerable end-to-end delay, resource constraints, multimedia source coding techniques, cross-layer coupling of functionality, and multimedia in-network processing [7].

These characteristics are handled either by modifying existing protocols in WSNs or by proposing new methodologies such as multi-radio multi-channel systems, switching between multiple channels, multipath routing, or mixtures of these methods [42]. In Table 2.1, we provide some popular applications of WMSNs, proposed multipath routing protocols, and the optimal solutions proposed for different application scenarios and environments. Generally, routing protocols for real-time applications impose severe demands on different QoS metrics such as low delay, high throughput, and high reliability [35]. Therefore, these characteristics, along with other research design issues such as security, connectivity, and coverage, have received considerable attention and are most likely to be considered in the different layers of the communication network protocol stack as shown in Figure 2.2 [43].

Energy efficiency is a common challenge for WMSNs, given the various energy-efficient mechanisms developed at the different layers of the network protocol stack [44]. For example, energy-aware routing protocols in which the available energy in each node can be used to select the optimal path while satisfying the QoS requirements has been proposed [44].

Table 2.1   Applications of WMSNs

Applications	Applications Scenario	Routing Protocol	Propose Solution
Smart environment	Power system, smart farming, water leakage, and forest fire detection	RTLD [8], Task allocation for real-time applications [9], RPAR [10], SONS [11]	Multipath multi-QoS constraints
		EQSR [12], RTRR [13], ICPSOA [14].	Multipath reliability constraints
		EADD [15]	Data-centric protocol
		SHRP [16]	On-demand fashion
Smart cities	General geographical tracking applications	SAR [17], SPEED [18,19], SPEED-EE [20], REAR [21], QuESt [22], THVR [23], Disjointed multipath routing [24], AMPMCR [25], IAMVD [26]	Multipath multi-QoS constraint
		MMSPEED [27], Re-InForM [28], NC-RMR [29], EERMR [30], DMRF [31], QoSMR [32], MCMP [33], ECMP [34], PMR [35], QoSNet [36]	Multipath reliability constraints
		DD [37]	Data-centric protocol
		RMDSR [38], TinyONDMR [39], RMRP [40]	On-demand Fashion
Security and emergency	Detection of gas, detection of people	Routing of high-priority packets in WSNs [41]	Multipath multi-QoS constraint

Multipath routing strategy has emerged as the technology of choice in WSNs, which can fulfill QoS metrics and networks constraints in most real-time applications [45]. The demands of multipath routing strategy combined with various performance demands dictated by different applications have led to the proposal of a number of new routing protocols to efficiently utilize the limited available resources

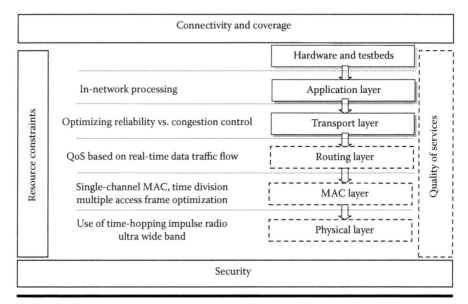

Figure 2.2 Research challenges at the various WMSNs layers of the communi-cation protocol stack. (Adapted from I. F. Akyildiz et al., *Computer Networks*, vol. 51, pp. 921–960, 2007.)

in addition to the features of multimedia data [46]. Moreover, the performance of traditional sensor nodes should be enhanced to keep with the fast-changing and realistic events in the real world [35].

Multipath routing operation is relevant for providing adequate network resources in various traffic conditions in order to fulfill these requirements [47]. Multipath routing techniques have been extensively used for improving the delivery of multimedia content, providing fault-tolerance routing and supporting QoS in networks from multi-hop local area networks (LANs), wireless area networks (WANs) and the Internet, and, finally, in ad hoc networks and WSNs [48]. Furthermore, the effectiveness of multipath routing strategies is essential to achieve high-quality network services and guaranteed QoS at high data rates. Consequently, the objective is to assign more loads to under-utilized multiple paths and fewer loads to over-committed paths so that uniform resource utilization of all available paths can be ensured to distribute network traffic [49,50]. Multipath routing protocol is cost-effective for heavy load balancing structures without dynamic traffic engineering [1,51]. To exploit multipath routing strategies, it is necessary to determine the number of available paths. Clearly, the number and quality of the selected paths dictate the performance of the multipath routing protocol. Thus, designing a multipath routing algorithm for multimedia applications is very challenging because of several characteristics and features of WMSNs. The breakdown of multipath routing schemes at different taxonomic levels and their features present a complete set of network routing protocols and highlight the key challenges of a communication routing protocol. With this broad picture in mind, we aim at assisting the reader in understanding what is essential and critical to QoS assurances with the current multipath strategies in WMSNs. In order to assist the readers, we provide in Table 2.2 a list of abbreviations, along with brief definitions as used throughout the chapter.

The rest of the chapter is organized as follows. Section 2.2 presents a review of related work. In Section 2.3, we present the system architecture and design issues of the routing algorithm, whereas in Section 2.4, we present the proposed high-level WSN multipath-routing protocol taxonomy. In Sections 2.6 and 2.7, state-of-the-art multipath routing techniques are presented. Section 2.8 provides a comprehensive analysis of major multipath routing approaches for WMSNs. In Section 2.9, some open issues and research directions are identified. Finally, conclusions are provided in Section 2.10.

2.2 Comparison with Related Survey Articles

Most of the existing surveys of QoS-enabled WSN routing protocols deal with application and architecture requirements along with their respective routing strategies [42,46,47,49]. However, there are additional criteria that should be emphasized when designing routing protocols for WMSNs. The most important of these are the lifetime of the multimedia sensor network and the overall QoS requirements of the respective real-time applications. We will review existing survey articles and highlight the gaps in those surveys as compared to the focus of this chapter.

Early surveys focused on general architectural issues and open research problems in the field of WSNs [42,52–60]. These surveys were based on the requirements of real-time applications and only cover certain aspects of QoS in WSNs, such as reliability, scalability, and ability to support real-time activities, which are not directly applicable to the more recent WMSNs. An introductory survey was presented in Reference 59 to discuss constraints and solutions related to each layer of the WSN stack, including sensor node hardware requirements. Open research challenges were presented for each layer. General research challenges in terms of operating systems, networking, and middleware protocol requirements to support various WSN requirements were presented in Reference 55. In Reference 42, routing protocols for WSNs were discussed and classified into two main categories: network structure and protocol operations. The authors outlined the design trade-offs between energy efficiency and communication overheads in the context of various routing paradigms. In Reference 59, the authors surveyed 27 WSN routing protocols that were current as of 2004 and presented a protocol classification based on multipath, query-based, negotiation-based, and QoS-based approaches. In addition, design issues such as network flow considerations and quality modeling were highlighted. A taxonomy for WSN routing protocol classification was developed in Reference 52, in which the system, network, operational, and objective models were used to classify various WSN routing protocols.

Table 2.2 Abbreviations

Abbreviation	Name
ADC	Analog-to-digital converter
AE	Available energy
AMPMCR	Adaptive multipath multi-constraint routing
AODV	Ad hoc on-demand vector
AOMDV	Ad hoc multipath demand vector
ARRCH	Adaptive reliable routing based on clustering hierarchy
BE	Best-effort
BF	Broadcasting flooding
BFS	Breadth first search
CC	Cell controller
CMOS	Complementary metal-oxide semiconductor
CMVT	Cross-layer multipath video transmission
CPU	Central processing unit
CR	Cognitive radio
CRSNs	Cognitive radio sensor networks
CSMA/CA	Channel sense multi-access/collision avoidance
DD	Directed diffusion
DFM	Duplication of forwarding messages
DMRF	Dynamical jumping real-time fault-tolerant
DPS	Dynamic packet state
DSA	Dynamic spectrum access
DSR	Dynamic source routing
EADD	Energy-aware directed diffusion
EBMR	Energy-balancing multipath routing
ECCs	Error correction codes
ECMP	Energy-constrained multipath
EDF	Earliest deadline first
EEPBC	End-to-End path battery cost
EERMR	Energy-efficient reliable multipath routing
EH-WSN	Energy-harvesting wireless sensor network
EQSR	Energy-efficient and QoS-based routing
EXT	Expected transmission count

(Continued)

Table 2.2 (*Continued*) Abbreviations

Abbreviation	Name
EYES	European Youth Environmental Sentinels
FECC	Forward error correcting codes
FIFO	First in, first out
GBR	Gadient-based routing
GPS	Global position system
HRT	Hard real-time
IAMVD	Interference-aware multipath routing for video
ICPSOA	Immune cooperative particle swarm optimization algorithm
IFS	Iterated function system
IoT	Internet of things
ISP	Internet service provider
LANs	Local area networks
LQEs	Link quality estimators
LQI	Link quality indicator
LTE-A	Long-term evolution-advanced
MAC	Medium access control
MANETs	Mobile ad hoc netwoks
MCMP	Multiple constraint multipath
MDP	Markov decision process
MIMO	Multiple input and multiple output
MIP	Mixed integer programming
MOGA	Multi-objective genetic algorithm
MP-DSR	Multipath dynamic source routing
NC-RMR	Network coding-reliable multipath routing
NSGA-II	Nondominated sorting genetic algorithm
OS	Operating system
PARSEC	Parallel simulation environment for a complex system
PMR	Partitioning multipath routing
PPP	Packet reception rate
QoS	Quality of service
QoSMR	Quality of services multipath routing
QoSNet	Quality of service network

(Continued)

Table 2.2 (*Continued*)   Abbreviations

Abbreviation	Name
QuESt	QoS-based energy-efficient routing
REAR	Real-time and energy-aware QoS routing
REER	Robust and energy efficient multipath routing
Re-InForM	Reliable information forwarding multiple paths
RMA	Rate monotonic algorithm
RMDSR	Robust multipath dynamic source routing
RMRP	Resilient multipath routing protocol
RPAR	Real-time power-aware routing
RRP	Rumor routing protocol
RSSI	Received signal strength indicator
RTLD	Real-time with load disturbed routing
RTRR	Real-time robust routing
SAR	Sequential assignment routing
SEEMR	Secure and energy-efficient multipath routing
SHRP	Shortest hierarchal routing protocol
SNR	Signal-to-noise ratio
soft-E2E	soft-end-to-end
SOCEE	Source Optimized Control with Energy Efficient
SONS	Self-organizing network survivability
SPEED	Stateless protocol for real-time communication in sensor networks
SPEED-EE	SPEED-energy efficient
SPIN	Sensor protocols for information via negotiation
SPIN-BC	SPIN-broadcast
SPIN-EC	SPIN-energy consumption
SPIN-PP	SPIN-point-to-point
SPIN-RL	SPIN-reliability
SRT	Soft real-time
TCP/IP	Transmission control protocol/Internet protocol
THVR	Two-hop velocity-based routing
TinyONDMR	Tiny optimal node-disjoint multipath routing
WAN	Wireless area networks
WFQ	Weighted fair queuing

(Continued)

Table 2.2 (*Continued*)   Abbreviations

Abbreviation	Name
WLAN	Wireless local area network
MDC	Multiple description coding
WMSN	Wireless multimedia sensor network
WSN	Wireless sensor network
XOR	Exclusive OR

Subsequent surveys considered additional issues such as energy efficiency [42], geographic location [61], network scalability [62], and cross-layer approaches [54]. In Reference 59, the advantages and performances of various approaches for real-time routing protocols and algorithms for WSNs were highlighted, and cross-layer design was introduced as a viable design approach. In Reference 54, a survey of various MAC and routing protocols for supporting real-time QoS for WSN in terms of reliability, data aggregation, and cross-layer protocol solutions were presented. In addition, trade-offs among different constraints in real-time applications, such as energy efficiency and delay performance, were highlighted. Energy-constrained routing protocols were studied in Reference 63, in which the taxonomy presented in Reference 42 was expanded to consider energy efficiency. Existing routing strategies were classified based on four main criteria: network structure, communication model, topology, and reliable routing approaches.

In Reference 61, position-based routing protocols were surveyed for WSNs. Flooding-based routing, curve-based routing, grid-based routing, and behavior-based routing were discussed. Metrics such as energy consumption, negotiation overheads, complexity, reliability, scalability, and multipath strategies were used for comparison. Scalability issues were also considered in Reference 64. Hierarchical routing strategies for large-scale WSNs were classified based on control overheads and energy consumption. Details of various protocols, as well as advantages and disadvantages in terms of the message complexity, memory requirements, localization, data aggregation, clustering algorithm, intra-cluster topology, cluster head selection, and multipath routing strategy were presented in Reference 64.

Another emerging trend in recent surveys is the focus on real-time multimedia data transmission in WSNs. In Reference 65, the authors surveyed issues related to supporting multimedia communication over WSNs. The design constraints and factors that influence multimedia delivery over WSNs, solutions appropriate for respective layers of the networking stack, along with their shortcomings and other major open research issues, were highlighted. Cross-layer designs for multimedia streaming were investigated in Reference 66. The paper discussed mechanisms for cross-layer optimization and outlined future research directions for each layer of the network stack. In Reference 43, architectures, algorithms, and protocols were proposed for the various layers of the networking protocol stack, as well as cross-layer designs for WMSNs along with an evaluation of the performance of existing hardware and test beds. Multipath routing techniques were introduced to address QoS requirements for WMSNs.

Energy efficiency and multipath routing approaches are the main focus of current research efforts in WMSNs. In Reference 47, energy-efficient routing strategies were introduced for WMSNs, including traditional schedulers (for example, rate monotonic algorithm [RMA] and earliest deadline first [EDF]), along with a discussion of the performance issues of each routing strategy, energy-efficient routing challenges for WMSNs, and limitations of current nonmultimedia routing strategies. The benefits of various multipath routing protocols for wireless sensor networks (WSNs) and their benefits were presented in References 66 and 67. The authors addressed the main characteristics of the multipath routing schemes and classified them according to their attributes. In Reference 46, multipath routing protocols were classified based on three main criteria: alternative path routing, reliable data transmission, and efficient resource utilization. The multipath routing protocol taxonomy in Reference 50 used path selection techniques and traffic distribution mechanisms to classify the surveyed protocols. The paper also discussed the suitability of the selected multipath routing protocols for meeting the performance requirements of various applications. Other considerations such as swarm intelligence-based routing, geographic awareness routing, and redundant traffic reduction based on the similarity of multimedia data sources from nearby locations were studied in Reference 7. However, the authors in Reference 7 classified the WMSN routing protocols according to the direction of sensor nodes equipped with multimedia devices. The classification depends on the routing system architecture and design issues for WMSNs.

The authors in Reference 46 classified a multipath routing taxonomy for WSNs into three categories, with emphasis on their advantages and disadvantages. The authors in Reference 68 surveyed multipath routing and provisioning in the Internet, as well as from the end-user and Internet service provider (ISP) perspective. However, our survey and the survey presented in Reference 46 have several major differences. First, our survey is more up-to-date and covers most existing multipath routing protocols of WMSNs, some of which are not covered in the previous work. Second, our survey focuses on the importance of particular multipath routing strategies for QoS assurances of real-time applications. Furthermore, we exploit mechanisms for the multipath routing strategy that aim not only at circumventing single-point failure but also to facilitate network provisioning. This investigation provides a comprehensive survey on both traditional data and more recent real-time multipath routing protocols.

Our survey presents an overall picture of QoS assurances in WMSNs via investigating and comparing advantages and disadvantages of existing multipath routing protocols. It should be pointed out that this chapter also provides an analysis of the classification of multipath routing protocols in terms of the exploitation and design issues for multipath strategies aimed not only at circumventing the multipath configuration but also at the effectiveness of network provision to network services and guarantees of QoS parameters. Table 2.3 provides a comparison between our survey and related survey articles available in the literature. To this end, a new classification

Table 2.3 A Comparison among Related Survey Articles

References	Considered Performance Metrics							Description
	Routing Strategies	Energy	Reliability	Scalability	Complexity	Topology	Security	
[52][a]								Used system, network, operational, and objective model metrics to classify various routing protocols.
[54][a]	✓	✓	✓	✓		✓		Surveyed MAC and routing that support real-time QoS for WSNs.
[55][a]	✓	✓	✓				✓	Surveyed general research challenges for supporting various WSN requirements.
[59][a]	✓	✓	✓	✓	✓	✓		Discussed node hardware requirements at each WSN stack layer.
[59][a]	✓	✓	✓	✓	✓	✓		Discussed routing design issues such as network flow and quality modeling.
[42][a]	✓	✓	✓	✓	✓	✓	✓	Discussed trade-offs between energy efficiency and communication overhead.
[61][a]	✓	✓	✓	✓	✓	✓		Surveyed position-based routing protocols.
[62][a]	✓	✓	✓	✓				Surveyed routing protocols considering scalability issues.

(Continued)

Table 2.3 (Continued) A Comparison among Related Survey Articles

	Considered Performance Metrics							
References	Routing Strategies	Energy	Reliability	Scalability	Complexity	Topology	Security	Description
[63][a]	✓	✓	✓	✓	✓	✓		Expanded taxonomy in Reference 42 to consider energy efficiency.
[64][a]	✓	✓	✓	✓	✓	✓		Surveyed hierarchical routing strategies for large-scale WSNs.
[65][b]		✓	✓					Surveyed issues related to supporting real-time multimedia streaming traffic without providing analysis of multipath mechanism.
[66][b]		✓	✓					Discussed mechanism for cross-layer design optimization but does not provide analysis of multipath mechanism in terms of exploitation and design.
[43][b]	✓	✓	✓	✓	✓	✓		Surveyed most routing protocols in WMSNs, along with introducing multipath routing mechanism, but does not provide any analysis of multipath mechanism in term of exploitation and design.

(Continued)

Table 2.3 (Continued)   A Comparison among Related Survey Articles

Considered Performance Metrics

References	Routing Strategies	Energy	Reliability	Scalability	Complexity	Topology	Security	Description
[47][b]	✓	✓	✓	✓	✓	✓		Introduced energy-efficient routing strategies.
[66,67][b]	✓	✓	✓	✓	✓	✓		Discussed the main characteristics of multipath routing schemes and provided classification according to their attributes.
[46][b]	✓	✓	✓	✓				Provided classification based on main criteria but does not focus on the importance of particular multipath routing strategies.
[47][b]	✓	✓	✓	✓	✓	✓		Introduced energy-efficient routing strategies.
[50][b]	✓	✓	✓	✓	✓			Used path selection techniques and traffic distribution mechanisms to classify the various surveyed protocols.
[7][b]	✓	✓	✓	✓	✓	✓		Provided classification for WMSN routing protocols according to the direction of sensor nodes equipped with multimedia devices.

(Continued)

Table 2.3 (Continued) A Comparison among Related Survey Articles

| References | Considered Performance Metrics | | | | | | | Description |
	Routing Strategies	Energy	Reliability	Scalability	Complexity	Topology	Security	
[68][b]	✓		✓	✓	✓	✓		Surveyed multipath routing and provisioning in the Internet, as well as from the end-user and ISP perspective.
Our	✓	✓	✓	✓	✓	✓		Provided an analysis of the classification of multipath routing protocols in terms of the exploitation and design issues for multipath strategies.

[a] Surveys based on the requirements of real-time applications and only cover certain aspects of QoS in WSNs. However, they can not be directly applied to WMSNs.

[b] Surveys based on classification of routing protocols. However, they did not cover the importance of particular multipath routing strategies for QoS assurances of real-time applications in WMSNs.

is proposed that classifies existing protocols according to the exploitation and selection of multiple paths. Moreover, a complete description with comparisons among different WMSN applications are presented in our survey.

2.3 Routing System Architecture and Design Issues

The extension of WMSNs has stretched the horizon of traditional WSNs. WMSNs allow monitoring and control of real-time information with low-cost and miniaturized sensor nodes. This brings different architectures, design goals, and constraints.

How can WMSNs reach to the derived reliable scalable architectural frame model to achieve QoS parameters with lower algorithmic complexity? Particularly, a routing strategy can be embedded into low-cost microprocessors to extend the lifetime of the sensor network without jeopardizing reliable and efficient communications. Generally, the architecture and design issues for routing protocols in WMSNs give rise to new challenges. The most dominant and challenging factors to achieve efficient communication in WMSNs are addressed in the following subsections.

2.3.1 Hard and Soft Real-Time Operation and Best-Effort for Resource Constraints

Real-time WMSNs consist of three types: hard and soft real-time systems and best-effort (BE) systems.

2.3.1.1 Hard Real-Time (HRT)

HRT applications involve a mechanism that operates on a preemptive and context-switching operating system (OS) with switching tasks in and out of execution modes based on the scheduled time slots to maintain real-time characteristics [56]. These applications might require an OS to deterministically delay different tasks. Traditional schedulers (for example, rate monotonic algorithm (RMA) and earliest deadline first [EDF]) might be suitable for such applications [56,57].

2.3.1.2 Soft Real-Time (SRT)

SRT applications involve a system that attempts to meet time constraints related to tasks, operations, and applications by applying system resources, including a high clock rate, fast processors, cache, and buses [56].

2.3.1.3 Best-Effort (BE)

BE applications are necessary to guarantee the best achievable performance; however, they may not be relatively focused on static measures such as central processing unit (CPU) allocation.

Designing routing protocols for WMSNs is influenced by many factors (such as the cost of hardware, network topology, power consumption, etc.), which all affect either probabilistic or deterministic end-to-end QoS metrics. When considering real-time support in WMSNs, energy efficiency should not be ignored, as energy consumption is the main constraint in WSNs. Therefore, there is often a trade-off among these constraints [69,70].

2.3.2 Energy Efficiency Model

WMSNs can be used in various applications spanning over an extended physical area. Therefore, for easier installation, the integration with a deployed wireless communication network module is a viable option. A large number of nodes are required for adequate coverage and collection of multimedia information for a prolonged duration. Typically, generating higher volumes of data traffic requires not only extensive processing but also higher transmission rates [47].

Basic components of a typical wireless multimedia node are illustrated in Figure 2.3 [71], including the sensor module responsible for capturing sensory information fed to a processing module. A sensor module consists of two sub-components: sensory devices for capturing a signal and an analog-to-digital converter (ADC). The main role of the processing module is to process the sensory data, encapsulation, and forwarding to other nodes or to a sink via a wireless module. Multimedia sensors provide audio and video sensing capabilities to make the sensor node operational in the WMSN environment. These multimedia extensions come

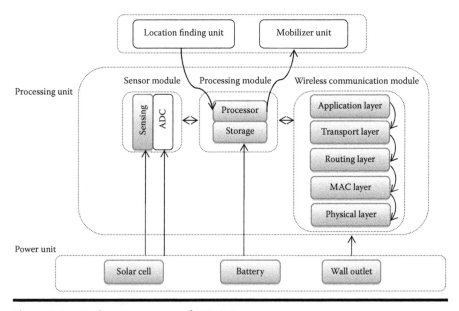

Figure 2.3 Basic components of WMSNs.

with depicting costs in the shaded unit in Figure 2.3. Because audio and video units are required, higher energy consumption is required, a larger space for random memory is expected, and a higher amount of bandwidth may be required due to the excessive multimedia content [71].

A major component in a sensor node is the power supply unit, which has limited power resources. Basically, the power unit consists of power managements (solar cell, battery, wall outlet, and/or capacitor) and an energy sink [72]. A special kind of sensor network, the energy harvesting-wireless sensor network (EH-WSN), in which each node is equipped with a rechargeable power supply instead of a traditional battery, has been reported [73–75]. Moreover, a battery is periodically recharged to keep the node continuously working rather than minimizing energy consumption. The node lifetime is defined by the time during which the energy level of the node is above a certain threshold that allows it to perform operations such as transmitting, receiving, or processing data [74]. Therefore, once the energy level drops below this threshold, the node should go to sleep to charge up; otherwise, the battery is considered dead and unable to continue the operations [74]. This means that it is desirable to know that different nodes harvest different amount of energy. However, this will be a tiresome task or might even be impossible since most WMSNs depend on deployment and environmental conditions that are needed to recharge the battery in order to get as much autonomy as possible [44]. Reducing or eliminating the problem of the limited lifetime will enable the network designers to develop the functionality of nodes by adding extra features and components [73].

A combination of energy-efficient mechanisms at the different layers in the protocol stack of WMSNs is used to minimize energy consumption and maximize the lifetime of the network. Thus, how could these routing protocols maximize the workload in the energy harvesting network as well as the lifetime of the networks? The answer to this question is to develop an energy-efficient routing protocol that is able to balance and optimally allocate available energy (AE) among active nodes. To improve energy-saving performance, each sensor node can be used to select the optimal paths based on the variation of environmental conditions [75].

Single-path and multipath routing mechanisms consider the most recent energy-efficient routing mechanisms according to the energy level of a particular node. Single-path approaches vary among different selected paths, thereby balancing the energy depletion among the nodes and extending the network lifetime. Multipath routing can achieve better balancing than single-path routing since multipath approaches can achieve load balancing and prevent congestion in the network by distributing the energy consumption among optimally selected paths [75].

2.3.3 QoS Modeling Requirements

End-to-end QoS requirements using a large number of mechanisms and algorithms have been pursued in different network stack layers to satisfy QoS parameters [42,59,60,70,71]. At the same time, different wireless network communications may impose specific constraints on supporting the QoS requirements, depending on their particular characteristics, challenges, and design requirements.

QoS support can be met by a plethora of approaches: for example, data-centric, protocol-centric, cross-layer optimization, cooperative, and distributed algorithms [47,76–78]. Our survey focuses on QoS requirements imposed by the applications on the network because of the proliferation of applications requiring some guarantee of services from the network and the performance of applications depends largely on the QoS assurances in WMSNs [78].

More particularly, how can the underlying network route the QoS-constrained data while efficiently utilizing network resources? The QoS requirements of WMSNs may be very different from those of WSNs in terms of delay, jitter, bandwidth, and energy consumption.

As a result, some WMSN applications rely on new QoS parameters that prefer measurement of the transmission of the sensor data in an efficient and effective manner [79]. Furthermore, network designers should be able to investigate which system architecture or mechanism can be exploited to provide better services than the BE services such as differentiated services (soft-QoS) and guaranteed services (hard-QoS) for the WMSNs applications. Some QoS requirements are briefly described in the following sub-sections.

2.3.3.1 Latencies

Ensuring stringent timeliness is essential for real-time video applications. Timeliness can be provided either on a guaranteed or BE basis according to the tolerance level of the multimedia application [47]. In addition to timeliness, there are several other factors that cause latencies in WMSNs, including in-network processing, transmission, and queuing delays [2,43,47]. Thus, reducing delays is a crucial task for real-time WMSNs applications. Generally, end-to-end latency can be classified into two classes: soft-latency bounded systems (deterministic constraint) and hard-latency bounded systems (predictive constraint) [47]. In hard-latency systems, the service cannot ensure meeting its deadline. Therefore, failure to do so is considered as a failure of the whole system, whereas the delay of a fraction of data traffic can be probabilistically guaranteed in soft-latency systems [47].

2.3.3.2 Bandwidth

Multimedia content, especially video streams, requires a high amount of bandwidth for transmission. In addition to multimedia transmission, some intermediate sensor nodes act only as relay nodes in a dynamic topology because of low transmission range and multi-hop communication. Therefore, the use of single-path routing strategy for transmission of multimedia may exhaust the energy of the path, resulting in network failure. Thus, a bandwidth constraint based on low-power transmission consumption must be handled by the multipath routing strategy [47].

Generally, QoS support becomes increasingly challenging because of the increasing desirability of connectivity and exchange of best-quality media information at any time, at any location, and in any manner. Therefore, designing effective routing strategies in WMSNs should be performed for QoS requirements for different classes of traffic patterns. This remains an open research area and a major challenge in the WMSNs field [80,81].

2.3.4 Sensor Network

Modeling large-scale sensor node deployment (such as for nodes communicating over wireless networks with lossy links and without infrastructure) is considered another challenge in WMSNs research [2,43,47]. The models can be classified either as deterministic or self-organizing (randomized). In the deterministic model, the sensor node is deployed, and packets are routed through predetermined paths. In the self-organizing model, sensor nodes are scattered randomly by creating an infrastructure in an ad hoc topology [82]. In terms of infrastructure distribution, the position of the sink, the cluster head, and the scope of the monitored area are crucial in terms of energy efficiency, performance, and real-time routing protocols. The architecture of WMSNs consists of three classes: single-layered, flat-layered, and homogeneous, as shown in Figure 2.4 [47].

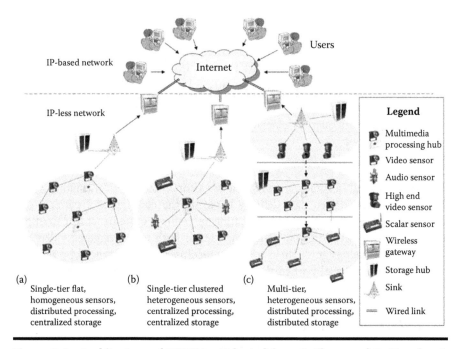

Figure 2.4 Architecture of WMSNs. (Adapted from S. Ehsan and B. Hamdaoui, *IEEE Communications Surveys & Tutorials,* **vol. 14, pp. 265–278, 2012.)**

WMSN architectures support not only heterogeneous nodes but also independent applications with different requirements. Thus, it is necessary to develop a flexible and hierarchical architecture that can accommodate the requirements of all applications in the same infrastructure [2,83]. Typically, heterogeneous sensory data are reported at different rates to some multimedia applications that may require a diverse mixture of additional QoS constraints [84].

2.3.5 Data Delivery Model

Data sensing and reporting models depend on the applications for both WSNs and WMSNs and the time criticality of data reporting. Because various applications have common requirements, it is sufficient to analyze each application by the classification of the data-delivery model. Such models are concerned with reporting data toward the sink along with their corresponding requirements. Generally, data sensing and reporting can be categorized into four basic traffic data delivery models or services: time-driven (continuous), event-driven, query-driven, and hybrid [80,85].

2.3.5.1 Continuous Time-Driven Delivery Model

This model is suitable for applications that require periodic dispatches of real-time data for monitoring, such as surveillance or reconnaissance, in which the sensor node constantly switches between the sensors and transmitters to sense the environment and transfer data in a periodic order.

2.3.5.2 Event-Driven and Query-Driven Models

These models consider the most commonly used applications in WMSNs. When a certain event occurs, the sensor nodes react immediately to the changes in the sensed attribute, or a query is generated by the base station [86].

2.3.5.3 Hybrid Model

This model is used in some networks that use a combination of continuous, event-driven, and query-driven data delivery based on the conditions of patterns carried out by the traffic and the classification of traffic requirements (for example, a surveillance application that periodically sends both event-triggered video and temperature). The design of the routing protocol is highly influenced by the data delivery model, especially when the design involves minimization of energy consumption and route stability [78].

2.3.6 Dynamic Network

Component-based architecture in the WSNs consists of three main components: the sensor node, sink, and monitored event [87]. Most researchers consider a stationary state of sensor nodes while designing the network architecture, whereas others consider the mobility status of sensor nodes and even the sink node to improve network conditions. In addition to energy consumption, real-time routing stability becomes another issue in routing the message to or from the mobile nodes and other QoS metrics [88]. The sensing/reporting can be either for dynamic or static events; for example, dynamic events are target detection/tracking applications, and forest monitoring for early fire prevention is a static event. Monitoring a static event allows networks to work in a reactive mode, simply generating traffic when reporting [59].

2.3.7 Reliability and Fault Tolerance

The concept of reliability concerns the ability to deliver multimedia data to the sink with minimum packet loss [47]. A different reliability constraint arises and needs to be imposed depending on the application demands that may require delivering the packets in a reliable transmission manner to the sink with timeliness and without loss of packets.

Similar to WSNs, the primary path in WMSNs may not often be able to support the sensed data because of broken links, among other reasons. Therefore, establishing the reliability of single-path and multipath routing protocols is challenging since multimedia applications generate high asymmetric traffic patterns, requiring high transmission rates and extensive processing. This challenge arises mainly because of the functional model of operational characteristics of routing protocols for data to reach to the sink. As long as WMSNs are used in various multimedia applications, this may generate a wide range of traffic patterns. The pattern of data traffic can be either single-hop or multi-hop [47,89]. The pattern can be divided according to the node density of the network or whether the network supports in-network processing, that is, the number of transmissions and reception of data packets processes in sensor nodes and/or whether any processing procedure is applied in the network, into the following patterns as depicted in Figure 2.5 [63].

2.3.7.1 Local Communication (Node-to-Node)

This type of communication has been proposed for ad hoc networks in which a node broadcasts its status to the nearest neighbor [80].

2.3.7.2 Point-to-Point (Node-Node)

This pattern is used to send an arbitrary node data packet to an arbitrarily directed node. This pattern is primarily used in wireless local area network (WLAN) environments. Point-to-point routing is not widely used for WSNs because building

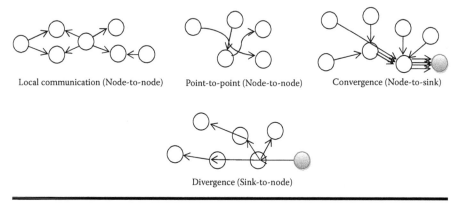

Local communication (Node-to-node)   Point-to-point (Node-to-node)   Convergence (Node-to-sink)

Divergence (Sink-to-node)

Figure 2.5   The pattern of data traffic in WMSNs. (Adapted from N. A. Pantazis et al., *IEEE Communications Surveys & Tutorials*, vol. 15, pp. 551–91, 2013.)

a routing tree does not scale well for sensor nodes that have only small data memory [90].

2.3.7.3 Convergence (Node-to-Sink)

This pattern needs to be supported to route responses back to the sink. Typically, this communication is defined as many-to-one (reverse multicast), in which multiple nodes or a single node respond for collection directly or indirectly to the sensed data packet and transmit it to the sink. This pattern is commonly used for data collection in WSNs and can also be unicast [90].

2.3.7.4 Divergence (Sink-to-Node)

This pattern needs support routing requests originated from the sink to multiple nodes. This is a general paradigm including other group-based routing concepts such as anycast (one-to-many), multicast, and manycast routing, implying that any node that has the requested data can respond to the query [70]. Particularly, a divergence traffic pattern is used when a sink node or base station sends control messages or queries to sensor nodes. Several techniques have been proposed to understand how different routing approaches respond when handling various requests in different network environments [42,91–98].

One of these techniques adopts a multipath routing strategy for performing fault-tolerant routing. The idea is to provide alternative or resilient routes for guaranteeing a reliable flow of transmission of data packets over multiple hops [99]: for example, sending copies of the same packets over multipaths to increase the probability that at least one of the copies reaches the sink with timelines.

2.3.8 Summary

Designing a routing scheme for real-time WMSNs is influenced by several factors, such as cost, transmission medium, network topology, power consumption, etc., which all affect either probabilistic or deterministic end-to-end QoS metrics guarantees, such as delay, jitter, and throughput. Moreover, WMSNs have specific requirements that are difficult to fulfill, such as traffic flow of large bursts of data at a high bit rate. These characteristics need to be addressed, taking into account hardware, bandwidth, and power limitations of the sensor nodes. Therefore, designing effective routing strategies that achieve all QoS requirements for different classes of multimedia data traffic remains a major challenge in WMSNs.

2.4 Routing Techniques in WMSNs Classification

A routing strategy is a key building block in a network protocol stack. It is a sub-component of the network layer in the sensor network stack layer model and is central to the proper functioning of any multi-hop communication system [63].

Routing strategy is defined in terms of the process of discovering and selecting paths in wireless networks from the target area that sends network traffic toward the sink node [42]. Therefore, the prime role of a routing protocol is to establish a path between the sources and sink node while keeping track of the path availability and facilitating successful transmissions of data along the selected paths [42]. In the following subsections, we will discuss the main classification of routing techniques of multipath routing protocols in the realm of WMSNs. Different criteria used to classify multipath routing protocols are given. Furthermore, we provide taxonomies for multipath routing protocols in WMSNs. Generally, there are three main phases of multipath routing: path discovery, path selection, and path maintenance. Once the paths are discovered, a routing protocol should decide how to select a path for sending data. Actually, the discovery, selection, and maintenance of paths depend on some node-specific and/or network-wide metrics such as QoS requirements, residual-energy budget along the selected path, and more [94]. Discovering and maintaining paths are also impacted by the data traffic, which is a major issue in designing the architecture of sensor nodes for multimedia applications. For example, the available buffer sizes, the limitation of resources (e.g., energy, memory, and processing power), QoS constraints, distribution of node density, their connectivity, and unexpected changes in node status during the duty cycling (e.g., inefficiency or failure) give rise to frequent and unforeseen topology alterations. Furthermore, these issues are also impacted by the fact that the common wireless channel is broadcast in nature, such as the control and corruption of data packets at the physical layer or collision of medium access control (MAC) protocols [100,101].

2.4.1 Designing Issues for Multipath Routing

Traditional routing approaches have been developed for cellular networks or wireless ad hoc networks and thus are not sufficient for WSNs, which are more demanding than other wireless networks [63]. Many new routing strategies have been proposed to solve routing problems in traditional WSNs [94–98]. The design of routing protocols for WMSNs, however, is still an open research area. In addition to the major issues of designing routing protocols in WSNs, there are new characteristics and constraints due to the nature of multimedia content that must be handled over the network such that routing protocols for WSNs are not applicable to WMSNs.

The most recent work seeks to handle these characteristics and their design challenges to solve the routing problem of streaming real-time multimedia content by modifying previous routing protocols in WSNs (for example, using multiple performance metrics to meet the additional QoS requirements) [42,102]. New solutions are proposed based on various new methodologies: for example, using multiradio, multichannel, or multiple-input and multiple-output (MIMO) systems, switching between multiple channels, selecting multipath routing, or a mixture of these methods [2,42]. Other approaches utilize optimization for cross-layer design between multimedia source coding techniques at the application layer and the routing layer to exploit optimal multipath selection or in-network processing [42]. In addition, the cross-layer design between the MAC layer and routing allows packet-level service differentiation or priority-based scheduling and more power-efficient routing mechanisms [50].

2.4.2 The Taxonomy of Multipath Routing Techniques

A plethora of research on taxonomies for mobile ad hoc networks (MANETs) and WSNs has reported the baseline model of routing protocols [7,103]. A routing protocol can be classified into one of four main parallel schemes, and each scheme is used to classify the routing protocol according to topology communication model, network structure, and reliable routing [59]. Some routing protocols adhere to QoS requirements in order to handle various types of multimedia and mixed traffic [104]. Designing multipath routing protocols for WMSNs requires additional research that extends the protocols proposed for WSNs [7]. Indeed, there are several ways to classify multipath routing protocols, depending on different parameters and factors such as reliability, latency, bandwidth, and load balancing [105]. One of these classifications is based on the number and type of QoS constraints of paths. The classes and sub-classes of classification are not mutually exclusive because many protocols belong to more than one class or sub-class [105]. Hence, the main division of taxonomy is based on equal assignment of different functionality to the sensor nodes.

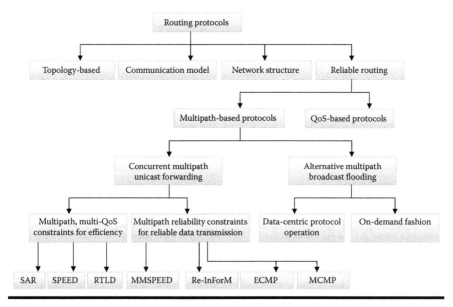

Figure 2.6   The taxonomy of multipath routing protocols.

Reliability has emerged as a factor to classify multipath routing protocols, since a reliable monitoring of the environment is very important in a variety of multimedia applications [106]. However, we use the reliability factor to classify the routing protocols according to the QoS requirements of multimedia applications. As depicted in Figure 2.6, multipath routing protocols can be classified into two main categories: concurrent multipath unicast forwarding and alternative multipath broadcast flooding. The first approach is based on sending multiple copies of the same data packets over multipaths, which increases the accuracy of various tracking in a variety of civilian and military applications [106]. The second approach is usually implemented by collecting or distributing information to all nodes in a wireless domain by performing a broadcast operation. In our classification, the decision to route data over multipaths depends on the mechanism of discovering and selecting a reliable path to meet certain objectives, such as congestion minimization, and application constraints (throughput, delay, and bandwidth), as shown in Figure 2.7. In particular, path selection mechanisms for traffic flow to reduce the overhead and complexity of the multipath routing scheme and to accommodate constraints on the selected paths are emphasized [68]. Therefore, the classification of multipath routing protocols is based on the way routing paths are established during the path discovery phase and the way routing paths are selected to distribute the traffic under specific constraints. To successfully achieve the application's requirements, multipath routing protocols may use the characteristics of the two methods of classification to govern the performance of WMSNs in terms of QoS parameters and energy efficiency.

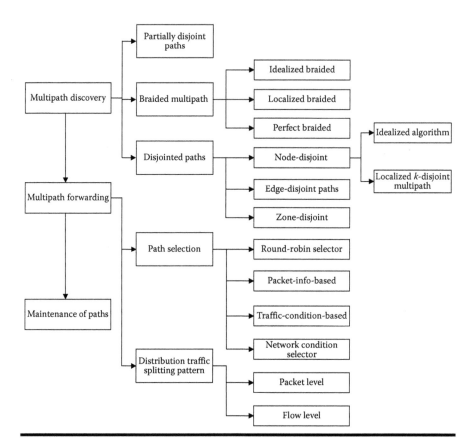

Figure 2.7   Taxonomy for the design of a multipath routing protocol.

Both taxonomies are complementary and suitable for meeting the requirements of WMSNs. They describe the determination of the availability of multiple paths as well as the distribution of traffic under multi-constraints and reliability. Figure 2.6 depicts the breakdown of the second level of the taxonomy of multipath routing protocols that will be enumerated in the subsequent sections.

2.4.3 Summary

Routing governs the performance of WMSNs in terms of QoS metrics. The design of a routing algorithm depends on several parameters that must accommodate multiple conflicting objectives and constraints, imposed by strategies on technologies and user requirements. Multipath routing is a promising strategy for achieving QoS metrics such as bandwidth, delay, and throughput as well as demands of the users. Several WMSN routing protocols have been proposed based on classifications that depend on different factors and on the users' requirements. We use the reliability factor to classify such routing protocols and to describe the benefits of multipath routing protocols according to QoS requirements in multimedia applications.

2.5 Multipath Routing Protocols: Challenges and Issues

Multimedia applications encompass monitoring of real-life events, which necessitates that efficient multipath routing mechanisms should be developed for the transmission of information while meeting QoS requirements [104]. The QoS requirements are expressed as a combination of QoS parameters of a multipath [107]. Thus, different multimedia applications have different QoS requirements, and multipath routing protocols in WSNs or WMSNs have their own unique advantages and disadvantages [94–96]. Consequently, it is a challenge to find or design an appropriate protocol or a class of protocols that fulfill all QoS requirements of an efficient routing protocol [35]. Furthermore, real-time multimedia applications encounter an additional challenge for energy-efficient multimedia processing [105]. Certainly, these challenges include optimal routing to meet the dynamic network constraints and application-specific QoS guarantees. In response to this challenge, particular QoS requirements require information of the current status of the network as well as resource constraints, since routing decisions are made based on this information [68]. For example, in certain applications, a multipath routing algorithm may select a set of node-disjoint paths with relaxation of node disjointness. This is referred to as a set of partially disjoint paths, or as a braided multipath that is link disjoint. For each node in the primary path, an optimally selected alternative path may not have any computed node in common. However, these kinds of alternative paths could potentially have comparable latency to the primary path and could therefore expend, more or less, the same amount of energy as the primary path [68]. Nevertheless, both link/node-disjoint paths improve reliability and offer more aggregate bandwidth than nondisjoint due to bottleneck link/node failure for nondisjoint paths, which negatively impacts the performance of multipath routing [106].

Many researchers seek to discover and select reliable paths by finding node-disjoint as well as link-disjoint paths. These algorithms do not take into consideration QoS parameters. However, successfully meeting certain objectives like QoS requirements and congestion minimization might require a global view of the network topology as well as its resources [108]. It should be noted that they are difficult to measure and generally can generate high overhead in networks [94], since resource availability information is periodically exchanged among nodes [68]. Other multipath routing protocols can aid in saving batteries by distributing network traffic uniformly among the sensor nodes; however, they increase the delay per packet transmission along the longer selected paths [109]. Many multipath constructing strategies have been proposed in the literature [94–96] to describe the performance of multipath routing protocols by the number and quality of selected paths. These constructions can be broadly classified into multipath discovery, forwarding, and maintenance [42]. Figure 2.7 illustrates the components and issues related to constructing multipath routes under the three categories discussed in this survey.

2.5.1 Multipath Discovery

As mentioned in Section 2.3.7, the pattern of data traffic in WSNs can be either single-hop or multihop. Therefore, the main task in multi-hop communication is to determine a set of intermediate nodes that should be selected [49].

Figure 2.8 illustrates the construction of partially disjoint paths and a braided multipath after the route discovery mechanism to create several paths from the source to the sink.

Obviously, designing a multipath routing protocol is more challenging than designing a single-path routing protocol due to the difficulty in finding the number of paths with a desired property in an effective and efficient manner. Furthermore, nondisjoint or single paths are easy to discover due to the absence of constraints on common nodes and links with any loop-free paths. Typically, this construction depends on many parameters to make the right routing decisions. Among these parameters is the extension of the unicast path to the multipath unicast path, which in turn depends on the amount of path disjointedness and is considered the main criterion used to utilize existing routing multipaths [46,94–96]. The performance of multipath discovery mechanisms depends on the number and quality of discovered paths, which in turn depends on the availability of network resources at intermediate paths, characteristics of paths, and QoS requirements [68]. There are

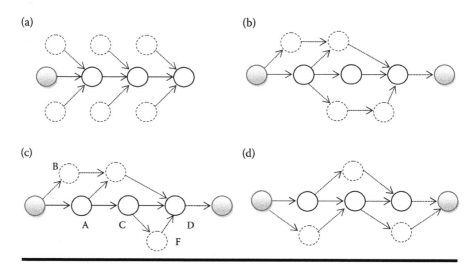

Figure 2.8 Construction of partially disjoint paths and braided multipath. (a) Fishbone structure, (b) Idealized, (c) Localized, (d) Perfect braided.

various multipath routing discovery mechanisms, as illustrated in Figure 2.8 and summarized below.

2.5.1.1 Partially Disjoint Paths

This mechanism is used to improve single-path routing protocols by providing a group of multiple alternative paths that are not simultaneously used in a large dense area [97,98]. Basically, the discovery mechanism takes advantage of the broadcast nature in which overhead packets of nodes are transmitted by their neighbors. Once the overheard packets arrive, a unicast route reply message is transmitted by each node's neighbor and records that neighbor as the next hop to the sink in its alternate route table [97,98]. Therefore, the multipath routing mechanism is able to discover a primary path with an alternate path that looks like a fishbone (see Figure 2.8a). The pattern of data traffic selects one of these discovered multiple paths at a time, whereas others are kept as a backup in case the path used becomes broken. However, if all discovered multiple paths are broken, then a new multipath discovery procedure is initiated. Typically, this mechanism is used to improve the reliability of data delivery and helps to reduce the overheard communication and end-to-end delay.

2.5.1.2 Braided Multipath

Also referred to as a "meshed multipath" [108], this mechanism is slightly different than the partially disjoint paths in that it increases resilience to node failure along with longer alternate node-disjoint paths. Because of some attractive resilience of partially disjoint path properties, more energy can be expended than that expended on the primary path.

Generally, there exist many possible definitions for data dissemination in the braided multipath mechanism. Therefore, a constructive definition for the braided multipath mechanism follows the directed diffusion paradigm that can be divided into three techniques [108]:

1. Idealized braided: For each node on the primary path, the gradient is computed to determine the preferred neighbor to find the best path from the source to the sink that does not contain that node. This results in a set of braided paths that lie either on the primary path or geographically close to the primary path as illustrated in Figure 2.8b.
2. Localized braided: The braided multipath technique relaxes the requirements like the idealized braided algorithm for node disjointedness of the complete multipath [108]. This technique uses two types of path reinforcement messages instead of finding a small number of alternate paths that depend on some localized techniques to construct braids at each node along the primary path. This technique can be briefly described as follows. The procedure initializes the sink to originate a primary path reinforcement message to the

next preferred neighbor B, as illustrated in Figure 2.8c. Once an intermediate node C has received the primary path reinforcement, it will transmit the primary path reinforcement to its next preferred neighbor, D. Thus, the constructed path traversed by the primary path reinforcement forms the primary path. In addition, the forwarding operation for each node C lies on the primary path, which also initiates a primary path reinforcement message to the next preferred neighbor, E. Once a node is not on the primary path, it will receive alternate path reinforcement, and it forwards to its next preferred neighbor. Otherwise, if a node is on the primary path, it stops the propagation of the alternate path reinforcement. The multipath structure formed by this technique is illustrated in Figure 2.8c.

3. Perfect braided: This technique achieves greater resilience of the primary braided path by generating a combination among the various alternate paths with failures on the primary path. For example, in Figure 2.8d, the number of distinct alternate paths is proportional to the nth Fibonacci number, where n is defined as the number of nodes on the primary braid path [108].

2.5.1.3 Disjoint Paths

This method is considered more attractive in many multipath routing applications because of the independence of the paths. Many multipath routing algorithms and protocols have been proposed to find disjoint paths in WSNs [35]. There are three types of disjoint paths: node-disjoint, edge-disjoint, and zone-disjoint. Clearly, both node-disjoint and edge-disjoint are the same and improve performance in terms of reliability [108].

1. Node-disjoint paths: This technique is of particular interest in many applications because of the independence and resilience provided by a number of alternate paths constructed within the primary path and also with each other. These discovered paths are unaffected by failures on the primary path but can potentially be less desirable than the primary path. This technique for constructing multipath-disjoint nodes can be divided, depending on the global knowledge of topology and network characteristics, into two types:

   a. Idealized algorithm: This algorithm uses two types of reinforcements. It initiates a low-rate scenario, as seen in Figure 2.9a, with a flooded network. Then, the sink has enough empirical information to determine a neighbor that can provide QoS metrics to send out primary-path reinforcement as illustrated in Figure 2.9b. Once the sink starts receiving data along with the primary path, the sink propagates the alternate path reinforcement to its next most preferred neighbor, A. This neighbor A continues the propagation in the same direction of the source to its most preferred neighbor, B. If B is already on the primary path between the source and sink, it sends a negative reinforcement to A (Figure 2.9c). Then A selects B as the next best preferred neighbor. Otherwise, B propagates the alternate path reinforcement to its most preferred neighbor (Figure 2.9d).

b. Localized *k*-disjoint multipath: The idealized algorithm mechanism can be extended to construct a *k*-disjoint multipath by sending out separated k-alternate path reinforcements from the sink to each next preferred neighbor. Figure 2.9e shows the difference between the localized algorithm and the idealized one, which takes in to consideration the performance of the best alternative path.

2. Edge-disjoint paths: This technique uses more specific, on-demand routing protocols. It is also called diversity injection and is used to find multiple disjoint paths between source and sink [108]. Generally, the procedure of route discovery of an on-demand routing protocol is initialized by broadcast route query messages by every node in the network. Once intermediates receive the messages, they only respond to the first received route query and discard the duplicate queries. Edge-disjoint paths attempt to reclaim the dropped information contained in the duplicate messages by recording, in a temporary query cache, the accumulated route information contained in all received route query messages. Because of the route query messages are only forwarded at each node, the received route messages at a node traverse various paths. By claiming the path information, a node acquires the reinforced route information and sends it back to the source.

3. Zone-disjoint paths: This technique defines an unselected path that has no shared nodes or edges with another path. Moreover, it cannot be within the interference range of the discovered paths, as shown in Figure 2.9f.

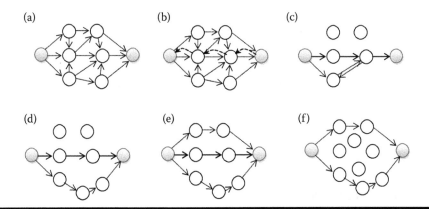

Figure 2.9 Illustration of the discovery of multipath mechanisms. (a) Low-rate samples, (b) Primary path, (c) Alternate path-negative reinforcement, (d) Alternative path P1, (e) Caveat, (f) Zone-disjoint paths.

4. Totally disjoint multipath: This includes the set of distinct paths that are zero-edge connected when concurrent data transmission takes place. Hence, the discovered multipath does not interfere, which is referred to as totally disjoint paths [108].

5. Maximally disjoint multipath: This refers to a set of node-disjoint paths that maximizes a disjoint characteristic among all possible paths while keeping common nodes at a minimum.

6. Radio disjoint multipath: The set of available paths with minimum radio interference, or the multiple noninterfering paths that are used to reduce the effect of interference between nodes as far as possible.

Due to QoS constraints, different resource limitations, and multimedia source coding techniques in WMSNs, the number of disjoined multipaths is considered as a fundamental challenge that should be taken into account when discovering a group of multipaths that may not be constructed of high-capacity paths.

The question is how can a forwarding path group be selected to discover multiple paths without having to consider energy consumption while still satisfying the QoS metrics along with the selected path? To answer such a question, it is necessary to define the problems associated with real-time support and reliability in wireless sensor applications.

Many routing protocols have been proposed [70]. One proposal is to select the number of optimal paths. Others just select the optimal path and maintain backup paths for the fault tolerance problem. Therefore, the operation of the selected paths is considered for better various QoS parameters [42].

2.5.2 Multipath Forwarding Models

A multipath forwarding model consists of two main components that are also important for load distribution: the selection path and distribution traffic splitting pattern, as illustrated in Figure 2.10 [1]. Various multipath forwarding models perform load traffic distribution in a different manner because of the difference in their internal functions of path selection and the distribution of the traffic splitting pattern, which may exhibit different advantages and shortcomings [1]. Multipath forwarding is among the contributions of this survey and is intended to assist the readers in understanding the different types of path selection mechanisms and traffic units. Figure 2.10 illustrates examples of various multipath forwarding mechanisms.

2.5.2.1 Path Selection

The selection of an adequate number of paths is considered a second important issue after the construction of multiple paths for data transmission purposes. Path selection for each multipath routing protocol is independently determined

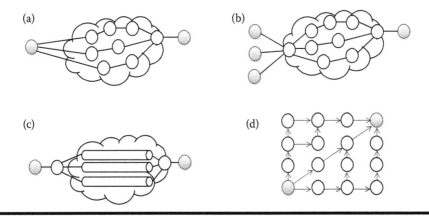

Figure 2.10   Various multipath forwarding configurations. (a) Multipath routing at source node with multiple intermediate nodes, (b) Multipath routing at multiple sources nodes with multiple intermediate nodes, (c) Multipath routing over multiple pipe-streaming links, (d) Multipath routing over wireless mesh, ad hoc, or sensor networks.

and should be selected to meet the performance demands of WMSN applications [1,46]. Most path selection schemes can be categorized into four types according to the purpose of designing a higher performance multipath routing protocol that selects a sufficient number of paths.

 a. Round-robin selector: The definition of this scheme is the successive traffic units that are sent across all parallel paths in a round-robin manner.
 b. Packet info-based: The path selection determines the packet identifier that is stored as information in the packet header of the arriving packet. The path selection is determined on the basis of the outcome of function identifier.
 c. Traffic condition-based selector: The selection of the path depends on the conditions of traffic distribution, such as traffic load, traffic rate, traffic volume, and number of data active flows.
 d. Network condition selector: The key to this selection is improving network performance such as throughput, delay, lifetime, etc. Hence, to improve resource utilization over the selected paths, the injected traffic rates of these selected paths are computed according to the path capacity, such as queue length.

2.5.2.2 Distribution Traffics Splitting Pattern

Suppose a group of multiple paths is selected, and then other issues arise, such as how a source node should transmit a packet. The source may split a packet into multiple segments and then transmit these segments by different multiple paths,

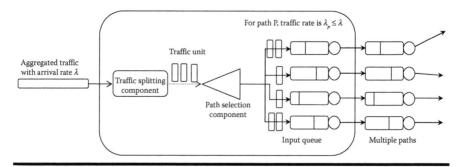

Figure 2.11 Multipath forwarding mechanism. (Adapted from N.S. Nandiraju et al., *IEEE Proceedings of the Conference International on Mobile Ad Hoc and Sensor Systems (MASS)*, pp. 741–746, 2006.)

or it may duplicate copies of packets using different multiple paths. There are two relevant aspects of an allocation strategy: granularity and scheduling [49,67]. The granularity specifies the smallest unit of information allocated to each path, as seen in Figure. 2.11 [110]:

    a. Per source-destination pair: based on using the same path to forward all traffic belonging to a certain pair of source and destination nodes.
    b. Per connection: based on allocating all traffic for the same connection to a single path.
    c. Per packet (packet-level): based on distributing the packets from multiple connections among the existing paths.
    d. Per segment (flow-level): based on splitting a packet into segments; each segment is forwarded using a different path.

2.5.3 Maintenance of Paths

In most multipath schemes, the source periodically floods the low-data rate alternate paths. To process, a maintenance phase reduces degraded multipath performance due to resource constraints with high dynamics of low-power wireless links of WSN. The main task of path maintenance in multipath routing protocols is to permit fast recovery from failures on the primary path. Three different situations are involved in this task:

    a. When the primary path has failed.
    b. When all discovered primary paths have failed.
    c. When some of the discovered primary paths have failed.

    In the first approach, the frequency of overall low-rate data flooding of paths to recover from failures on the primary path causes latency and a high overhead. Moreover, the discontinuity of a route rediscovery mechanism until failure of all active paths may increase energy consumption and reduce network perfor-mance. The third approach is more interesting than the others since it represents a trade-off between energy expenses and the likelihood of total multipath failure [46,111].

2.5.4 Summary

The design of multipath routing protocols for WMSNs is crucial for the provi-sion of QoS. This opportunity has posed new challenges since many trade-offs need to be considered. Firstly is ensuring reliable and energy-efficient end-to-end multipaths to transmit data. Secondly, the large number of available paths combined with multiple paths for each increases the cost of route processing for path selection. It should be noted that the processing and exchanging of information among the different layers of WMSNs impedes the performance of communication.

The challenges associated with design issues for WMSN routing protocols come from all sorts of functionalities of application areas, such as military, health, and other application facilities to meet the QoS requirements. We can group these chal-lenges under two main categories: reliability and timeliness assurance of multipath routing protocols.

2.6 Concurrent Multipath Unicast Forwarding

Unicast path routing protocol covers the construction of a sequence of single, efficient quality links from the source to the sink, possibly over multiple hops, eventually providing the significance of the cost of a path. It is the same as the flat routing protocol operation, in which each node collaborates with others to perform the task of constructing an optimal path that is called a simple flood operation.

The extension of unicast path to multipath unicast depends on the mechanism of construction, selection, and distribution of the optimal *n*-paths between the source and the sink. This extension is classified into three main operations: dis-covery, selection, and maintenance. In the operation of discovering paths from the source toward the destination, it is common to use dissemination approaches that can be easily adopted in WSNs.

2.6.1 Multipath Multi-QoS Constraints for Efficient Resource Allocation

Many researchers have focused on the network layer, as the multipath routing pro-tocol has always played a crucial role in supporting these routing metrics. Many available protocols tackled these constraints, specifically QoS parameters with an energy routing metric [112]. The following sub-sections address traditional end-to-end QoS assurances in various specific traffic patterns.

2.6.1.1 Principal Protocols of Multipath Multi-QoS Constraints

Concerning multipath multi-QoS constraints, there are several principal protocols utilized to achieve the multi-QoS constraints for efficient resource allocation.

Sequential Assignment Routing (SAR) was the first routing protocol developed for WSNs [17] in which QoS issues for making a routing decision were considered based on three factors: energy conservation, QoS parameters, and level of packet priority in the traffic flow. These traffic types were applied through a given flow for each data packet with a constant priority, and they remain unchanged until they reach the final destination. SAR uses a table-driven multipath approach that satisfies the QoS parameters, energy consumption, and fault tolerance. The disadvantage of SAR is that the creation mechanism of the multipath causes additional node energy depletion. Thus, it is not suitable for multimedia transmission.

The Stateless Protocol for Real-time Communications in Sensor Networks (SPEED) is considered the first protocol that envisioned soft real-time requirements under specified constraints [18,19]. Its localization/geographical protocol provides guarantees to support QoS parameters for soft real-time traffic. It supports three types of data: unicast, multicast, and anycast. SPEED maintains a desired delivery speed across the network through a novel combination of the nondeterministic QoS awareness of geographic forwarding and feedback control. This combination of MAC (single-hop) layer and network (multi-hop) layer adopts a cross-layer approach that improves the end-to-end delay transmission time and provides a good response to congestion and voids. However, the disadvantage is in the prolonged lifetime of the sensor node that is achieved by only the reduction of control packets and geographic routing without consideration of other energy metrics during the routing operation.

SPEED-Energy Efficient (SPEED-EE) improves the network lifetime of the SPEED protocol by employing an optimization algorithm [20]. The protocol provides a protection for nodes with less energy to avoid their discharge and to provide the nodes with more energy in order to increase the lifetime of the network. The residual energy is calculated based on the algorithm reported in References 113 and 114, whereas the delay is computed as in the SPEED protocol [18]. The disadvantage of this protocol is the adaptation of the selected path, which may increase the overhead problem between the sensor nodes over a large-scale area.

Real-time and Energy-Aware QoS Routing (REAR) routing protocol has been applied to WMSNs by using an advanced Dijkstra algorithm to evaluate the structure of a multipath mechanism and chooses the neighboring distance from node i and node j amid all paths to send real data [21]. To reduce queue delay for real-time event packets, a classifier queue model is used for each node to deal with real-time and ordinary data and to balance the network life-cycle. The disadvantage of this protocol is the complexity overhead in creating the multipath algorithm because of the parameters that are associated with each path.

Real-time with Load Disturbed Routing (RTLD) computes the optimal forwarding hop based on three metrics: link packet reception rate (PRR), residual power of the sensor's battery, and packet velocity per single-hop [8,115]. The routing protocol consists of four functional components: power management, neighborhood management, location management, and routing management. Each component response to the specific management of the sensor node is shown in Figure 2.12 [103]. The routing management computes the optimal forwarding choice based on three parameters to choose the optimal forwarding: speed of moving packet through the hop, packet received ratio (PRR) and remaining power for every single-hop neighbor, and the delay from the source node to the single-hop neighbor. The disadvantage of this protocol is the elaboration of the energy consumption profile for the diverse levels of the duty cycle of the sensor network, which is used to the derive the trade-off between energy conservation and QoS.

Energy-Aware Delay-Constrained Routing is an energy-efficient routing scheme for WSNs that ensures low delay for data delivery in the network [116]. Considering the transmission power, sensor's energy, and the end-to-end delay as constraints, this approach sets up energy-aware multi-hop data paths. End-to-end delay is achieved by using a weighted fair-queuing (WFQ) scheduling technique in each sensor. In addition, leaky-bucket traffic-regulation techniques are utilized to regulate the incoming traffic from the sources and separate the real-time traffic from the non-real-time traffic by applying two different queues in each node. The disadvantage of this protocol is the WFQ technique, which utilizes a highly complex approach for packet scheduling. Moreover, there is a need for tag computation and tag sorting at line speeds. This requirement presents a bottleneck problem that is not suitable for multimedia transmission.

QoS-Based Energy-Efficient Routing (QuESt) was proposed in Reference 22 by using a multi-objective genetic algorithm that determines a set of nondominated near-optimal paths in a flat network topology (nonclustered). The algorithm satisfies application-specific QoS parameters, end-to-end delay, bandwidth requirements,

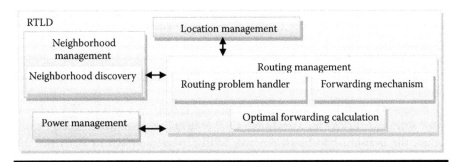

Figure 2.12   Block diagram of RTLD routing protocol.

and energy consumption in WSNs. To satisfy different QoS parameters, constraints for QoS-routing problems were mapped into a constrained Steiner tree problem. Moreover, the proposed algorithm uses Euclidean distance to present two randomly selected paths from the population. Unfortunately, the Euclidean distance of the Steiner tree problem has proven to be a nondeterministic polynomial-hard problem (NP-hard), with no currently known polynomial time algorithms.

Routing of High-Priority Packets was designed in Reference 41 to suggest a real-time adaptive capability to identify unusual events in the presence of routine data traffic. The classification of the packet depends on whether the data in the packet differs significantly or not from the contents of the previous packet from the same source. The classification setting may be normal (routine), that is, including the same data from previous packets. Otherwise, the packet's content is very different from that of the previous packets, at which point the priority bit will be set to one to indicate an unusual state. The routing function decides the routes for the data packet with high priority. That is, if the incoming rate of the unusual packets is measured by a source node that is higher than a given traffic threshold value θ_U, then the source node chooses to forward all high-priority packets to the destination along the optimal QoS path. Otherwise, the high-priority packet rate drops down the traffic measurements window, and the nodes re-route the packets to their destination along shorter paths. The disadvantage of this protocol is the unsuitability of the mechanism of re-routing at a large scale, given the complexity of the time slice of the re-routing operation. In addition, the complexity overhead of the shortest-path algorithm during the creation/ updating of the routing table at each round-robin slice is increased.

Real-time Power-Aware Routing (RPAR) was proposed in Reference 10 to address the main challenges of support for real-time communication. This protocol improves the number of packets delivered before deadlines with low energy cost. The authors assumed that all sensor nodes are stationary and know their location via the global position system (GPS). The proposed protocol was designed to work with existing simple channel sense multi-access/collision avoidance (CSMA/CA) asynchronous MAC protocols such as the X-MAC or B-MAC protocols. RPAR consists of four components: a dynamic velocity assignment policy, a delay estimator, a forwarding policy, and a neighborhood manager. The disadvantage of this protocol is that the velocity increases exponentially for each packet as the one-hop distance increases but sharply decreases because of the degradation in link quality. Unfortunately, increasing power transmission causes more contending nodes and higher collisions when nodes receive two overlapping packets. The location management requires the additional cost of GPS hardware to identify the node's location.

Enhancing Real-time Delivery with Two-Hop Information or Two-Hop Velocity-based Routing (THVR) for a real-time QoS protocol with improved energy utilization, the over-one-hop-based protocol SPEED was proposed in Reference 23. The idea of using the information of two-hop routing is a trade-off between performance improvement and complexity. The information exchange between two hops starts by sending the message "HELLO" to two intermediate sensor nodes. Then, each successive node sends messages to all of its neighbors informing about the information of their one-hop neighbors. The disadvantage of this protocol is that it is limited only to two-hop information for optimal decision forwarding. The increase in complexity and overhead with an increasing number of hops is high. Moreover, whenever the number of hops has increased, it will give $O(n^4)$ time for improving the performance in energy consumption, where n is the number of nodes in the network.

The Self-Organizing Network Survivability (SONS) routing protocol was proposed in Reference 11 to monitor forest fires. It is an extension of Reference 117 using hierarchical real-time QoS routing to provide survivability and data reliability. Multiple hop-spanning trees were used to achieve a large coverage area. The weakness of SONS is the synchronization phase during the spanning tree formation needed for dynamic power management, which can experience a large number of collisions in areas with high node density.

Tasks Allocation for Real-time Applications in Hierarchical Heterogeneous Networks has been proposed in Reference 9. The main objective is to find an optimal allocation that minimizes the overall energy consumption while meeting the application's deadline. The task allocation algorithm exploits the divide-and-conquer algorithm to efficiently solve the scheduling problem for large-scale allocation applications through four methods: task partition, deadline distribution, scheduling, and re-scheduling. A Markov decision process (MDP) is used as the decision-modeling tool for solving the optimal scheduling problem. MDP helps to find the optimal policy for all possible states in the Markov diagram. The disadvantage of the protocol is its time complexity in both scheduling and re-scheduling and the computation of optimal policy for task allocation.

Disjointed Multipath Routing for Real-time Data: A hybrid k-disjoint multipath routing protocol was proposed in Reference 24 for WMSNs. This protocol is based on a combination of the strengths of two transmission schemes: the Zigbee and Bluetooth. To effectively transmit multimedia data, the data of the source node are split through Zigbee to be transmitted to several neighboring nodes localized in a Bluetooth scatternet. Then, the protocol minimizes path setting overhead through a competition-based nonoverlapping multipath for data transmission. The splitting mechanism has a first-in, first-out (FIFO) pattern of transmission sequencing on individual paths to combine original multimedia data without sorting in the sink.

Some researchers proposed several multipath routing protocols for traffic load balancing and repartition [118–121]. However, most of these protocols did not optimize the traffic load balance and repartition against all selected paths and even according to the resource limitations of WMSNs [2]. Therefore, according to the characteristics of WMSNs with the expected network lifetime, the authors in Reference 122 proposed a node-disjoint design that emphasizes limited energy to maximize the data-gathering performance. The proposed algorithm is composed of two phases. The first phase determines the set of multiple paths and finds the node-disjoint path. The second phase selects paths from the set of multiple paths based on load balancing to balance the load in the network. Consequently, the bandwidth is utilized efficiently via a new congestion control message. A gradual-increase strategy based on paths and a gradual-increase strategy based on flows are introduced. This routing algorithm is a proactive routing protocol because it picks alternate node disjoint paths from the routing table. However, it fails to note how the multiple node disjoint paths are identified between the source and the sink. The traffic is equally distributed among the multiple paths. This mechanism is feasible for resource constrained networks such as multimedia sensor networks.

The multipath routing protocol has been explored to increase the probability of reliable data delivery that is appropriate for WMSNs. Therefore, the **Energy-balancing Multipath Routing (EBMR)** in Reference 123 is proposed as a node-disjoint routing protocol based on the node's residual energy and distance between the node and the sink. The protocol is composed of two phases. The first is to find the multiple node-disjoint routing paths from the source to the sink while excluding the source node. The second is the next hop selection mechanism allows common sensor nodes to more evenly consume energy. This leads to an increase in the performance of EBMR in energy balancing and network lifetime, respectively, by calculating the cost function value. More precisely, the current node selects its neighboring nodes, whose cost function value is as small as the next hop node. The cost function considers not only the distance between the selected node and the sink but also the neighbor node's residual energy. However, when the neighbor node's energy is less than the mean residual energy of the neighbor set, its cost function is added. Otherwise, its cost function value is subtracted.

In Reference 25, the authors presented an inter-cluster based on an **Adaptive Multipath Multi-Constraint Routing (AMPMCR)** to decrease the loss rate, energy consumption, and delay with the help of link quality. The cost function between the links is calculated by using the delay, loss rate, and remaining energy. Initially, a cluster-based architecture is constructed for the WSN in which the cluster head is determined based on energy level and capacity. However, a cluster-based architecture is considered where inter-cluster routing is performed based on a weight cost function. During routing, the cluster head in each level of clustering selects its higher-level cluster with a cost function value. The nodes that are not part of routing are placed in sleep mode to save energy. Simulation results display several advantages in terms of a lower delay and higher delivery ratio than other QoS protocols and display a decreased number of dropped packets.

The major challenge in WMSNs is the transmission of video. Because of the large data size, video transmission consumes the node's energy. Moreover, conventional routing finds a relatively shorter path between the source and the sink; however, this approach is unviable for multimedia applications. To solve the problem, a cross-layer multipath video transmission (CMVT) routing scheme reported in Reference 124 presents the idea of differentiated service and multipath routing operating in both the application layer and network layer. Through collaboration between the application and network layers, CMVT combines the advantages of differentiated service and multipath routing to provide different paths for different applications. The combination starts from the application layer, which is responsible for video gathering and encoding frames and then distinguishing them according to their importance by marking them with different tags. Once they arrive at the network layer that was considered as the core of CMVT, the routing protocol assigns them to different paths, and all of the important frames are guaranteed with reliable paths. CMVT has an obvious superiority in media transmission, especially in large-scale WMSNs.

Interference information can be used not only to determine the routing path conditions but also to establish the new routing path. Reference 26 proposes a novel **Interference-Aware Multipath Routing for Video Delivery (IAMVD)** in the realm of WMSNs. The proposed routing algorithm discovers two node-disjoint, interference-minimized paths for a single pair of the source nodes and the sink. Then, the routing mechanism is performed without any special hardware support for localization, making it practical for resource-constrained WMSNs. IAMVD uses a sleeping mechanism to create a block area of nodes preventing additional energy consumption. Moreover, the effects of different QoS requirements for multi-priority packets is considered in the process of path construction and video transmission by dedicating various multiple paths to various priorities.

2.6.1.2 Discussion

There are many research challenges when using WSNs in real-world interactions because of constraints imposed by energy consumption and unreliable wireless channels. Moreover, other QoS constraints that may also be important for particular applications such as security and density deployment overlap should also be considered at the different layers of the protocol stack. Although a few studies focusing on supporting QoS constraints in WMSNs have been reported [24–26,123] there are several interesting studies regarding supporting the QoS of real-time applications [8,10,11,17,19–23,41,115,116]. These studies proposed different mechanisms and protocols for the different layers of the network protocol stack to enforce QoS metrics, since access to information from lower layers and from higher layers may be required. Nearly, all of these mechanisms have been developed and tested with various network simulators, such as network simulator-2 and OMNeT++ [17,18,21].

Because supporting multi-constrained QoS parameters in WMSNs is affected by design choices at the various layers (such as physical, medium access, and network), the multipath strategy provides an integrated approach to manage QoS. The multipath strategy is in the right direction when sensor operations are involved in the real world, as long as the strategy provides reliable communication and load balancing for deadlines associated with end-to-end routing. The multipath routing strategy should be designed with a trade-off between energy efficiency and multi-constrained QoS parameters to guarantee an efficient usage of the amount of energy available at each sensor node and to achieve an efficient resource allocation. The design of multipath routing protocols depends on the load distribution forwarding models, which involves dynamic traffic engineering to exploit the multiple paths discovered that facilitate network provision to facilitate high-quality network services and to guarantee multi-constrained QoS. Because it provides load balancing, without dynamic loading, traffic engineering may heavily incur use of the multipath routing. Therefore, designing a multipath routing strategy configuration for forwarding multimedia data in a sensor network should be engineered by traffic splitting and the mechanism of path selection in heterogeneous traffic environments. This approach is used by several QoS routing algorithms to compute multi-constrained multiple paths by trading between the optimality of the multipaths and the complexity of the load distribution forwarding model [8,10, 11,21–23,41,115,116].

A multipath mechanism is pursued to develop routing solutions that also comply with QoS constraints in the design methods. Tables 2.4 and 2.5 depict a comparison among the aforementioned multipath unicast forwarding routing protocols for real-time issues on the basis of QoS constraints, data delivery, network architecture, and other parameters.

Most reported routing protocols did not address the computational complexity and lack an analytical load distribution model for the network performance because of the multi-constrained QoS parameters that are faced with time complexity and/or space complexity.

This section concludes that a multipath routing approach should formulate the problem of finding a path subject to additive/multiplicative multi-constraints of QoS routing in an analytical way. The model then provides a routing protocol for dynamic traffic engineering that implements capacity provision based on the distribution of end-to-end QoS parameters using different optimization approaches.

2.6.2 Multipath Reliability Constraint for Reliable Data Transmission

In many contemporary applications, both real-time and reliability characteristics are essential. Reliability is a real challenge in WSNs, as the nodes are exposed to many failure modes because of hardware failure, energy depletion, and communication link errors [2,42]. Furthermore, each failure mode decreases the performance of the network.

In WMSNs, reliability exists at many layers, such as a software layer, hardware layer, communication link layer, applications layer, and even at the network layer. At the network layer, reliability implies that the routing protocol provides fault tolerance by increasing the probability of a data packet sent by a source node, via alternatives of selected paths, arriving at the destination sink with minimum packet loss [46].

Table 2.4 Comparison of Multipath Unicast Forwarding QoS Constraints for WMSNs

Routing Protocol	Scalability	Energy Efficiency	Reliability	Network Topology	Data Delivery Model	Network Dynamic	Resources Reservation	Data Aggregation	QoS	Localization
SAR [85]	Restricted	High	Restricted	Flat	Query-driven	+	+	+	+	
SPEED [86,87]	Restricted	Low	Restricted	Geo.	Query-driven	+			+	
SPEED-EE [88]	High	High	Restricted	Flat	Hybrid	+			+	
REAR [91]	Restricted	High	Restricted	Flat	Geographic	+			+	
RTLD [92],[93][a]	High	High	Restricted	Flat	Query-driven	+	+		+	
EADR [94]	High	High	Restricted	Hierarch	Query-driven	+	+	+	+	
QuESt [95]	Moderate	High	Moderate	Tree	Query-driven		+	+	+	
RRR [96]	Restricted	Moderate	High	Tree	Event-driven	+	+	+		
RPAR [97]	High	High	High	Flat	Query-driven	+	+		+	+
THVR [98]	Restricted	High	High	Flat	Query-driven	+	+		+	
SONS [99]	High	High	Restricted	Hierarch.	Event-driven	+	+	+	+	+
TAP [101]	Restricted	Moderate	Restricted	Flat	Query-driven	+	+			

(Continued)

Table 2.4 (Continued)   Comparison of Multipath Unicast Forwarding QoS Constraints for WMSNs

Routing Protocol	Scalability	Energy Efficiency	Reliability	Network Topology	Data Delivery Model	Network Dynamic	Resources Reservation	Data Aggregation	QoS	Localization
Disjointed multipath routing protocol [102]	Restricted	High	Restricted	Ring	Hybrid	+	+		+	+
EBMR [108]	Restricted	High	Moderate	Flat	Event-driven	+	+	+	+	+
AMPMCR [109]	Restricted	High	Restricted	Flat	Hybrid	+	+	+	+	+
IAMVD [111]	Restricted	High	Moderate	Flat	Hybrid	+	+	+	+	+

[a] All protocols satisfy the soft real-time QoS constraints; however, the RTLD protocol satisfies both hard and soft real-time QoS constraints.

Table 2.5 Summary of Multipath Mechanism for Unicast Forwarding QoS Constraints for WMSNs

Routing Protocol	Path Disjointedness	Number of Paths	Path Selection	Traffic Distribution	Path Maintenance
SAR [17]	Partially disjoint	Based on the desired reliable tree	Sink	Traffic condition-based selector	Discover new path
SPEED [18,19]	Partially disjoint	Based on the desired reliable discovered paths	Source and intermediate nodes	Packet info-based	Discover new path
SPEED-EE [20]	Partially disjoint	Based on the desired reliable discovered paths	Source and intermediate nodes	Packet info-based	Discover new path
REAR [21]	Braided multipath	Based on the desired reliable discovered paths	Sink	Traffic condition-based selector	Discover new path
RTLD [8,115]	Localized braided	Based on the desired reliable discovered paths	Source and intermediate nodes	Network condition selector	Discover new path
EADR [116]	Path disjointedness	Based on the desired reliable discovered paths	Sink	Traffic condition-based selector	Discover new path
QuESt [22]	Disjoint paths	Based on the desired reliable tree	Sink	Traffic condition-based selector	Discover new path
RRR [41]	Node-disjoint paths	Based on the desired reliable routing function	Source	Packet-info-based	Discover new path

(Continued)

Table 2.5 (Continued) Summary of Multipath Mechanism for Unicast Forwarding QoS Constraints for WMSNs

Routing Protocol	Path Disjointedness	Number of Paths	Path Selection	Traffic Distribution	Path Maintenance
RPAR [10]	Perfect braided	Based on the desired reliable discovered path	Source	Traffic condition-based selector	Discover new path
THVR [23]	Node-disjoint	Based on the desired reliable routing discovered path	Source and intermediate nodes	Traffic condition-based selector	Discover new path
SONS [11]	Idealized algorithm-node-disjoint	Based on the desired reliable tree	Source and intermediate nodes	Network condition selector	Discover new path
TAP [9]	Node-disjoint	Based on the desired reliable tree	Sink	Traffic condition-based selector	Discover new path
Disjointed Multipath routing protocol [24]	Node-disjoint	Based on the desired reliable tree	Sink	Traffic condition-based selector	Discover new path
EBMR [123]	Node-disjoint	Based on the desired reliable tree	Sink	Traffic condition-based selector	Discover new path
AMPMCR [25]	Node-disjoint	Based on the desired reliable tree	Sink	Traffic condition-based selector	Discover new path
IAMVD [26]	Node-disjoint	Based on the desired reliable tree	Sink	Traffic condition-based selector	Discover new path

Any real-time application over a WSN should consider resource constraints, global time varying network performance, and communication reliability [125].

2.6.2.1 Principal Protocols for Multipath Reliability Constraint

As far as multipath reliability constraint is concerned, there are several principal protocols that are utilized to achieve reliable data transmission.

Multipath Multi-SPEED (MMSPEED) [27] is the extension of the SPEED protocol. The MMSPEED routing protocol provides several improvements and modifications to guarantee QoS parameters in WSNs over the network and MAC layers. First, a packet data delivery mechanism that provides QoS differentiation in two quality domains, timeliness and reliability, is proposed. Second, an end-to-end QoS reliability according to the local decisions for each intermediate sensor node is proposed. This end-to-end QoS reliability is considered an important property for large-scale sensor networks. The timeliness domain provides a classifier from various traffic types to classify the multiple network-wide speed options for processing the packet. Classifying works properly by choosing the speed options depends on the data packet deadlines, as shown in Figure 2.13. The disadvantage is that the protocol is not compatible with large-scale, long-distance transmission because the time estimation of reachabilities of the nodes increases exponentially with respect to the hop distance between the current neighborhood and the sink.

The Reliable Information Forwarding Multiple Paths (Re-InForM) protocol in Reference 28 employs a probabilistic flooding to deliver an information awareness packet and service at the desired priority levels of reliability at a proportional cost for sensor networks through dynamic and randomized multipath forwarding mechanisms to forward multiple copies of the same information packet in multipath routing according to the decision of the source to route the packets toward the sink. The routing mechanism is based on the local knowledge of network conditions such as channel errors, hops counting to sink, and out-degree. Information on network conditions is stored in the head of the packets header without requiring any data caching at any sensor node by using the dynamic packet state (DPS) method that causes an increase in the probability of information delivery. The disadvantage of the protocol

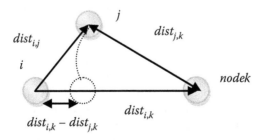

Figure 2.13   Progress speed between two intermediate nodes.

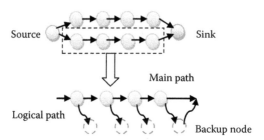

Figure 2.14   NC-RMR topology.

is the duplication of packets that might cause a high cost of energy consumption and occupying a useful channel bandwidth utilization.

The Network Coding-Reliable Multipath Routing (NC-RMR) protocol for WSNs was presented in Reference 29. The NC-RMR protocol employs computations of paths and next-hop node selection as in the Re-InForM protocol, with differences from Re-InForM being shown in Figure 2.14. The first difference is avoiding the redundancy of packet copies. NC-RMR applies the network coding mechanism in delivering packets through a multipath from the source to the sink. Second, to increase the level of reliability, the NC-RMR protocol employs a hop-to-hop mechanism to establish a disjointed and braided multipath routing protocol. Third, the NC-RMR protocol includes load balancing by the braided multipath and optimal next hop-to-hop node selection. Fourth, the security analysis is based on the coding division of an initial message into partitions and then combines the coding partitions at the intermediate nodes. The disadvantage is that although node-disjoint multipath routing conserves energy, the path selection may yield more hops to reach the destination. Conversely, NC-RMR saves around 67% in maintenance and complexity of overhead problems, which provides 50% higher resilience to node failure.

The Energy-Efficient Reliable Multipath Routing (EERMR) protocol for data gathering was proposed in Reference 30 for WSNs. This protocol achieves a reliable transmission with low energy consumption by utilizing the energy of intermediate nodes to gather and distribute data to the sink based on their requirements. The authors propose a heuristic algorithm, the energy-efficient multipath tree construction, to maximize the minimum residual energy among the nodes. The packet dispersion mechanism is used for splitting data packets at the source into segments and distributes them on multiple parallel paths to reduce the packet loss. The packets are re-assembled and aggregated at the destination. The disadvantage of the protocol is that it is suitable only for large-scale areas, as the greedy algorithm suffers from high computational complexity. Moreover, the greedy algorithm fails sometimes to find the optimal local solution; therefore, it might produce the worst possible solution.

The Dynamical Jumping Real-Time Fault-Tolerant (DMRF) routing protocol was proposed in Reference 31. The DMRF protocol focuses on reliability at the

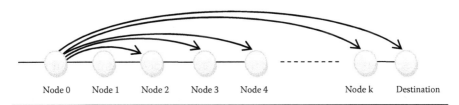

Figure 2.15   Jumping transmission.

network layer and attempts to reduce the effect of the sensor node failure from the empty congestion area in the network domain to guarantee real-time performance. DMRF routing assumes that the node uses the remaining transmission time of each sensor for the packets, and then the state of the candidate node in the forwarding path dynamically selects the next hop through a constrained transmission interval. Discarding data packets due to sensor node failure might cause wastage of transmission energy. To avoid energy waste, the DMRF protocol proposes that the transmission mode be set to jump in a way to reduce the transmission time as shown in Figure 2.15. Based on the feedback from downstream nodes, every node adjusts the jumping probabilities to raise the ratio of successful transmissions. The disadvantage of the DMRF protocol is the dynamic jumping, which continues in a large-scale area in order to select the next hop that might lead to an increase in energy cost. Therefore, the whole network lifetime cannot be prolonged.

Self-Organizing and Collaborative Energy-Efficient Sensor Networks developed from a study that was part of the European Education Youth Environment Sentinels (EYES) project (IST-2001-34734) on self-organizing and collaborative energy-efficient sensor networks [117]. The proposal of an approach that ensures gathering of information at the sink as sensor nodes may not be available during the procedure as reported in Reference 126. The algorithm begins with discovering k-multiple paths from the source to the destination. The calculation of these paths is based on the reputation coefficient of the nodes using the Bernoulli-distributed parameter. If a node has a low reputation coefficient, that is, if it frequently fails to route the packets, then it should be avoided in the routing scheme. The reputation coefficient increases with each successfully routed packet. Forward error correcting codes (FECC) are used to split the original data packet into sub-packets and allow the reconstruction of the original message by adding the redundancy.

The Quality of Service Multipath Routing (QoSMR) protocol proposed in Reference 32 considers only reliability as the constraint without tackling the delay constraint. The QoSMR protocol is an extension of secure and energy-efficient multipath routing (SEEMR) [127]. The QoSMR protocol starts with the same adoption mechanisms as SEEMR with topology construction. To guarantee the reliability constraint, QoSMR protocol uses two mechanisms to reduce collision that occurs when the source nodes send data packets to the sink node at the same time. First, a query seeks specific data through a simple flood, then the sink finds the paths after receiving a reply from the sensor nodes that match the flood query. Second, if no disjoint paths are available, then the sink schedules the data transmission of each source node. To avoid having the sharing nodes instantaneously receive plenty of packets from multiple source nodes, the revised breadth first search (BFS) algorithm is used for searching and re-scheduling individual paths for other different sensor nodes if there are no disjoint multipaths available. The disadvantage of this protocol is the space and time complexity for topology during search and re-scheduling. The space complexity is expressed as $O(|V|)$, where V is the number of available multiple paths, while $(|V|^2)$ is the time required for searching independent braided multipaths for different source nodes.

The *Multi-Constrained Multipath (MCMP)* [33] focuses on providing soft-end-to-end (soft-E2E) QoS (reliability and delay) in multipath routing by using an analytical formulation. In MCMP, the end-to-end soft QoS problem has been formulated rigorously using stochastic programming. The interpretation of the scheme is based on a combination of resources of multiple paths for traffic flow. The delay constraint, denoted as d_1, associated with data packet delivery is considered. If there is no single feasible path for satisfying reliability, r_1 multipath routing can improve it. The probabilistic delay-reliability constraints problem $\forall path \in P(Source, Destination)$ is expressed as

$$min \sum_{j=1}^{P} x_j \tag{2.1}$$

Then the first constraint is

$$x_j * d_j \leq Delay \tag{2.2}$$

The second constraint is

$$r = 1 - \prod_{j=1}^{P} 1 - x_j * r_j \geq Reliability \tag{2.3}$$

where $x_j = 0$ or 1 for all $j = 1, 2, ...,$ and $P x_j$ is defined as the decision variable, whether the selected route is j or not. Many other constraints are assembled to increase the energy efficiency of multipath routing. For example, power consumption, maximizing throughput, etc. By focusing on fulfilling the requirements of the soft-E2E QoS for a long feasible path raises many questions. First, "How can the reliability achieved by a subset of paths be quantified?" Second, "How can the energy-efficient path be chosen subject to delay constraints?" These questions are considered as nonlinear programming problems. In an attempt to solve these problems, a model based on the probability exploration to provide soft-E2E QoS under multiple constraints for a multipath routing protocol was proposed in Reference 35. The authors suggested a distributable manner as the main requirements to achieve the high level of soft-E2E QoS of the selection path. The partition was obtained from the hop requirements for both the additive delay and multiplicative reliability, which is formulated as

$$L_i^d = \frac{BoundedDelay - Delaynode_i}{Hopcount} \tag{2.4}$$

where *Hopcount* is a counter of hops from the source to the sink, *Delaynode$_i$* is the value of the delay engrossed for processing data at *node$_i$*, and *BoundedDelay* is the definition of the total delay from the source to the sink. Reliability formation is formed as

$$L_i^d = \sqrt[Hopcount]{Re_i} \tag{2.5}$$

where *Re$_i$* is the reliability requirement assigned to the path through *node$_i$*. The focus is on the definition of the nonlinear programming problem and where it should be started. Some problems depend only on the source node based on information on the end-to-end solution, while other problems depend approximately on the intermediate node based on hop information. Both problems attempt to reduce the complexity of constraints. Thus, the authors estimated the probability of the constraints according to the one-tailed version of Chebyshev's Inequality. The proposed protocol was evaluated through the parallel simulation environment for a complex system (PARSEC) that supports parallel discrete-event simulation capability. The disadvantage of the protocol is the dependence on linear and nonlinear programming to solve optimization problems. These methods are inefficient for multi-constrained optimization problems owing to the large number of calculations required for an efficient search of an optimal multipath solution.

In the **Energy-Constrained Multipath (ECMP)**, energy, delay, and reliability become competitive constraints in WSNs, raising a trade-off between the single-path and multipath routing deployments, where minimizing energy and delay and maximizing reliability is at stake. The authors in Reference 34 extended the models proposed in Reference 33 into a model referred to as ECMP by building upon geo-spatial energy propagation to formulate QoS routing in WSNs. The MCMP model proposes an arbitrary selection link with a random choice if it is the optimal selection for minimizing energy consumption. Thus, the ECMP model tries to find a subset from a set of sensor nodes with lower expected energy transmission, while achieving the QoS requirements. The ECMP searches for the subset of multipath from the source to the sink, which satisfies the QoS and the total energy of transmission requirements. The disadvantage of this protocol is the geo-spatial energy propagation approach, which depends on the Pythaorean theorem that leads to the overhead messages control problem.

An integrated QoS network for a routing protocol in large-scale WSNs [36] is a new routing protocol (quality of service network [QoSNet]) that increases the network lifetime by using the percolation theory while satisfying the hard-QoS constraints. The system model considered sensor nodes deployed inside a two-dimensional area formed by a grid network architecture. The approximate location of each sensor node's coordinates is obtained through triangulation or multi-lateralization. The proposed system model presents the same method as the fractal theory iterated function system (IFS), which has the ability to work in mobile indoor/outdoor environments but with costly deployments. The IFS is compatible with a large-scale sensor area to discretize the wireless sensor area into constructor cells, and each cell is equipped with a cell controller (CC) that supports a simple sensor node whose only role consists of performing data for-warding. The explanation behind the proposed QoSNet model is related to the goals. First, each cell is equipped with a CC responsible for managing the cell QoS-path selection operation and, second, supports the management to integrate powerful entities as shown in Figure 2.16, which, for a square, creates two layers of sensor nodes: the lower represents the entire network, while the upper is com-posed exclusively of CCs. Data is forwarded through typical scenarios when an abnormal event is detected. It is necessary to forward through a group of feasible paths via the intermediate sensor nodes until reaching its destination. For this pur-pose, each sensor node maintains a table for records and data-processing reporting session. The authors added a new constraint definition, namely end-to-end path battery cost (EEPBC), to extend the lifetime of each individual sensor node and to prevent early failures.

An *Energy-Efficient and QoS-Based Routing (EQSR)* protocol was proposed in Reference 12 for WSNs. EQSR is an extension of the robust and energy-efficient multipath routing (REER) protocol [128]. REER was used to examine traffic allocation by two methods. Firstly, a single-path among the discovered paths was

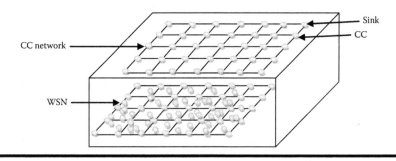

Figure 2.16 Two-layer overview of the WSNs.

used to transfer the message. Secondly, the transmitted message is split into several equal-size segments including exclusive OR (XOR) based on error correction codes (ECCs), and then these segments are simultaneously transmitted across multiple paths to increase the probability of receiving enough portions of the packet at the destination without increasing the delay.

The authors extended their protocol to restore failures in the node and to balance the load by splitting the traffic across a group of node-disjoint paths to balance the energy consumption among the sensor nodes. The reliability of data delivery increases by sending the data with ECC computation data.

EQSR utilizes available buffer size, signal-to-noise ratio (SNR), and residual energy to predict the next hop during the path construction phase. During the path discovery phase, the link cost function was used to examine the decision for the recent link performance; then, predictions of path stability may be made. EQSR employs the WFQ presented in Reference 129 to handle both real-time and non-real-time traffic. Two queues are used: a high-priority queue for real-time traffic and a FIFO queue for non-real-time traffic. The EQSR protocol has two disadvantages: WFQ is a highly complex approach for packet scheduling, and overhead complexity might occur during the decision on path stability.

The *Real-time Robust Routing (RTRR)* protocol was proposed in Reference 13 for fire emergencies in indoor building scenarios. Each sensor node is defined by four state messages: safe, lowsafe, infire, and unsafe. In the RTRR protocol, a simple delay value is used, and the forwarding decision must satisfy the slack where it is less than the average delay.

Adaptive reliable routing based on cluster hierarchy for WMSNs is necessary to adapt to the differences in energy consumption in WSNs and WMSNs. WMSNs consume energy in the sensing status and in processing multimedia data, while WSNs consume energy only in the communication links. These differences are considered in the **Adaptive Reliable Routing based on Clustering Hierarchy (ARRCH)** [130] to handle challenges faced in WMSN routing, such as energy consumption, limited computing power, memory availability, and QoS constraints. The authors presented a self-adaptive power allocation mechanism as a metric to meet the reliability requirement.

An energy-efficient QoS routing in a two-tiered WSN was proposed by a **Multi-Objective Genetic Algorithm (MOGA)**. A protocol for energy-efficient QoS-routing based on a nondominated sorting genetic algorithm (NSGA-II) for WMSNs was reported in Reference 131. In NSGA-II, the ranks of chromosomes are presented in the same manner as in MOGA. Optimizing a particular objective function for the end-to-end delay, reliability, and energy consumption were considered in Reference 22. The main objective is to find a solution that satisfies the best trade-off among the three objectives, called Pareto Optimal.

An Immune Cooperative Particle Swarm Optimization Algorithm (ICPSOA) for fault-tolerant routing optimization in heterogeneous WNSs uses a hybrid routing scheme to calculate and maintain k-disjoint multipaths from source to sink [14]. ICPSOA was presented as a model to solve the fault-tolerant optimization routing for sensor nodes that are densely distributed in a heterogeneous wireless environment. ICPSOA is an intelligent swarm algorithm that provides a faster way to recover the *k*-disjoint multipath from failure. The authors used a simple form of fault tolerance proposed in Reference 132, which is defined as sub-trees of a modeling-directed connected graph. Usually, the construction of topological heterogeneous WSNs consists of two types of sensor equipment, arranged in two layers. The lower layer is formed by traditional sensor nodes with restricted resources, which respond for any task, such as processing, transmission, and sensing data. The second layer consists of macro-nodes with more capabilities in energy, processing, and storage, which have the responsibility of decision routing. The metric for selecting the optimal path from the *k*-disjoint path spanning in the sub-trees connected graph is defined as the ratio of the total energy consumption of the valid path from the sensor node to the macro-node/root to the summation of the energy consumption path function, communication delay function, and distance function between all nodes in the sub-tree. ICPSOA abrogates the problem of converging to an undesired optimal solution, although ICPSOA derives better diversity. The increasing diversity depends on developing a searching mechanism inspired by the immune cooperative, which defines each particle as an antibody to generate a new search in the space population. ICPSOA provides accuracy in finding the optimal solution by jumping out to the local optimal solution, with minimum energy consumption, and shorting end-to-end delay for packets delivery. The algorithm suffers from a lack of robustness for link failure in the network and its computational complexity.

The *Partitioning Multipath Routing (PMR)* protocol was proposed in Reference 35 for WMSNs. The authors proposed a routing metric to determine the optimal path and to select the intermediate sensor nodes for routing packets from a source to a sink. The metric prioritizes the sensor nodes according to the link quality based on a generic routing protocol. The mechanism of path selection is based on defining critical parameters to control the adaptive switching of hop-by-hop QoS routing protocols by using mixed integer programming (MIP). The embedded criteria for each objective function related to decision constraints are used to select the path from the source to the sink.

2.6.2.2 Discussion

In addition to the traditional QoS requirements such as delay and throughput, the diverse reliability constraint should also be considered in WMSNs. Reliable communication is defined as another challenge facing WSNs because of some constraints, such as the characteristics of time-varying channels for low-power wireless links, interferences, and dynamic network topology. As previously mentioned, the multipath routing strategy appears to be an essential feature for supporting QoS in WMSNs. Thus, this strategy improves the chances of data transmission and reception during transmission to the sink without any interruption, even in the case of path failure by using different approaches to providing reliable data transmission in alternative paths. There are various approaches to provide a reliable transmission. The first, transmits multiple copies of data packets, as long as the discovered multipaths exist, to ensure recovery from several path failures. The second method uses an erasure coding technique in which each source adds some additional information to the original packets, and then the data packets are distributed over multipaths.

The multipath configurations, which can be established in several different ways, as shown in Figure 2.9, influence the performance benefits achieved to guarantee integration with other networks. Many existing network topologies with various environments are depicted in Figure 2.10. To provide a reliable data transmission in each situation, multipaths are generated that may be established by using a totally different approach.

Tables 2.6 and 2.7 summarize and provide comparisons of the aforementioned multipath unicast forwarding routing protocols for reliability issues based on QoS constraints, data delivery, network architecture, and other parameters.

2.7 Alternative Multipath Broadcast Flooding

As mentioned before, one goal of any WSN is to prolong the lifetime of the sensor node in the network. To achieve this goal, it is necessary to construct an energy-efficient path without imposing stringent QoS requirements from the source node to the sink node, possibly over multiple hops. This is usually performed by collecting or distributing information to all nodes in the wireless domain by a broadcasting operation.

Owing to various mechanisms in the broadcasting flooding (BF) operation and its low-rate flooding in discovering the entirety of available multipaths, multipath BF sub-taxonomy can be classified into two partitions, depending on the main operation: indicator-based and indicator-free [133]. In the indicator-based partition, there is always an initialization phase in which an indicator-generation algorithm is applied. According to the algorithm, every node generates an indicator to help determine the routes.

There are different types of indicators that further categorize this type of operation into subclasses, namely the data-centric protocol, where indicators are built for sensors in a setup stage. They then follow those indicators to make decisions while routing. In the indicator-free partition, the algorithm has no initialization phase and the packets are transmitted in an on-demand or random fashion, which is not applicable for WMSNs [133]. It is essential to describe the meaning of the flooding operation and grossing problem and highlight the basic idea before describing a few proposed protocols of the data-centric protocol and on-demand fashion.

Table 2.6 Comparison of Multipath Unicast Forwarding Reliability Constraints for WMSNs

Routing Protocol	Scalability	Energy Efficiency	Reliability	Network Topology	Data Delivery Model	Network Dynamic	Resources Reservation	Data Aggregation	QoS	Localization	Mobility
MMSPEED [27]	Restricted	Low	Restricted	Flat	Geographic	+			+	+	
Re-InForM [28]	High	Low	High	Flat	Query-driven	+			+	+	
NC-RMR [29]	Restricted	Moderate	High	Flat	Query-driven	+					
EERMR [30]	Restricted	High	Restricted	Flat	Query-driven			+	+		
DMRF [31]	Restricted	Moderate	High	Flat	Query-driven	+	+	+	+		
SOCEE [126]	Restricted	Moderate	High	Flat	Query-driven	+	+	+	+		
QoSMR [32]	Restricted	High	High	Flat	Query-driven		+	+	+		
MCMP [33]	Restricted	Low	High	Flat	Event-driven	+	+	+			
ECMP [34]	Restricted	High	High	Flat	Event-driven	+	+	+	+	+	
QoSNET [36][a]	High	High	High	Mesh	Event-driven	+	+	+	+	+	+
EQSR [12]	Restricted	High	High	Flat	Event-driven		+	+	+		
RTRR [13]	Restricted	Moderate	High	Flat	Event-driven	+		+	+		
ARRCH [130]	High	High	High	Star	Event-driven	+	+	+			
NSGA-II [131]	Moderate	High	Moderate	Tree	Query-driven		+	+	+		
ICPSOA [14]	Restricted	High	High	Flat	Query-driven		+	+			
PMR [35]	Restricted	High	High	Flat	Event-driven	+	+	+	+		

[a] All protocols are satisfying the soft real-time QoS constraints, except the QoSNET protocol, which is satisfying the hard real-time QoS constraints.

Table 2.7 Summary of Multipath Mechanisms for Unicast Forwarding Reliability Constraint for WMSNs

Routing Protocol	Path Disjointedness	Number of Paths	Path Selection	Traffic Distribution	Reliability Mechanism
MMSPEED [27]	Partially disjoint	Based on the desired reliable tree	Sink	Packet condition info-based	Copying the original packets
Re-InForM [28]	Link disjoint	Based on the desired reliable discovered paths	Source	Packet info-based	Multiple copies of the original packets
NC-RMR [29]	Partially disjoint	Based on the desired reliable discovered paths	Source and intermediate nodes	Packet info-based	Multiple copies of each packet
EERMR [30]	Partially disjoint	Based on the desired reliable discovered paths	Source and intermediate nodes	Packet info-based	Multiple copies of each packet
DMRF [31]	Braided multipath	Based on the desired reliable discovered paths	Sink	Traffic condition-based selector	Per packet hopping
SOCEE [126]	k-disjoint paths	Based on the desired reliable discovered paths	Sink	Traffic condition-based selector	Per packet hopping
QoSMR [32]	Partially disjoint	Based on the desired reliable discovered path	Sink	Traffic condition-based selector	Per packet for each path
MCMP [33]	Partially disjoint	Based on the desired reliable path	Intermediate nodes	Packet info-based	Multiple copies of each packet

(Continued)

Table 2.7 (Continued) Summary of Multipath Mechanisms for Unicast Forwarding Reliability Constraint for WMSNs

Routing Protocol	Path Disjointedness	Number of Paths	Path Selection	Traffic Distribution	Reliability Mechanism
ECMP [34]	Partially disjoint	Based on the desired reliable and energy consumption	Intermediate nodes	Packet info-based	Multiple copies of each packet
QoSNET [36]	k-disjoint paths	Based on the probability of discovered path	Source and intermediate nodes	Traffic condition-based selector	Multiple copies of each packet
EQSR [12]	Node-disjoint	Based on the probability of successful data transmission	Source	Distributed traffic splitting	Multiple copies of each packet
RTRR [13]	Idealized algorithm node-disjoint	Based on the desired reliable tree	Source and intermediate nodes	Network condition-based selector	Per packet, each message
ARRCH [130]	Node-disjoint	Based on the desired reliable tree	Sink	Traffic condition-based selector	Copying the original packets
NSGA-II [131]	Node-disjoint	Based on the desired reliable tree	Sink	Traffic condition-based selector	Copying the original packets
ICPSOA [14]	k-disjoint paths	Based on the desired reliable tree	Sink	Traffic condition-based selector	Copying the original packets
PMR [35]	Partially disjoint	Based on the desired reliable and energy consumption	Source and intermediate nodes	Traffic condition-based selector	Copying the original packets

2.7.1 Data-Centric Protocol Operation

Flooding is considered as the multi-hop routing algorithm for wireless communication environments [134]. Whenever a node receives a packet, the node broadcasts this packet to all of its neighbors. Broadcasts continue until all of the nodes in the network receive the packet. As a result, a packet can be flooded through the whole network. The flooding operation can be controlled by limiting the re-broadcasts until the packet reaches the destination or the maximum number of hops has been reached. Flooding is classified as a reactive operation, and its implementation is straightforward. The flooding enjoys the advantages of simplicity, since a node does not require neighborhood information and does not require costly topology maintenance or complex route-discovery algorithms [134].

2.7.2 Data-Centric Protocol Problem

An example of a realistic scenario is forest fire detection. The information packet that contains the 100°F temperature is more important than sensing the temperature on a spring day with 60°F. The high temperature might cause a forest fire, and thus it should reach the sink with high reliability, low latency, high bandwidth, and a high packet delivery ratio. BF can be used to deliver important packets with high reliability, but the overhead incurred is significant and consumes energy of the sensor node because of the duplicate messages during the broadcasting operation. The deployed mechanism (i.e., the network density) and the number of hops from source to sink might cause very low delivery of a packet to the sink because it is not feasible to assign global identities to each node.

2.7.2.1 Principal Protocols

Sensor Protocols for Information via Negotiation (SPIN) [135] was the first protocol in a data-centric family using meta-data negotiation-based, data-centric, and time-driven flooding operation protocols. SPIN created a family of different routing operations classified as point-to-point (SPIN-PP), energy consumption (SPIN-EC), broadcast networks (SPIN-BC), and reliability (SPIN-RL). The rumor routing protocol (RRP) [136] was defined as a classic flooding operation that can be described as an event flooding. Query flooding describes the process used to propagate queries to all nodes in the network when no localization information is available to steer the query toward the appropriate sensors. The gradient-based routing (GBR) [137] is considered the third data-centric protocol in which a gradient is determined based on the number of hops to the sink.

Directed Diffusion (DD) is proposed in Reference 37 for WSNs and is considered as one of the flat routing protocols. DD paved the way for many protocols based on DD or following similar concepts [15,59]. Generally, DD consists of several elements: interests, data messages, gradients, and reinforcements, as seen in Figure 2.17.

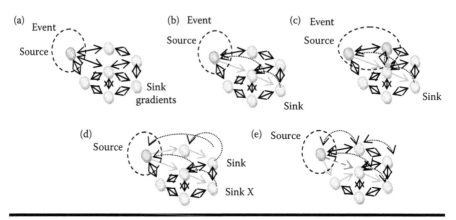

Figure 2.17   The construction of the DD routing protocol: (a) Gradient establishment, (b) Reinforcement, (c) Multiple sources, (d) Multiple sinks, (e) Repair.

An interesting message is a query defined by a list of attribute-value pairs, such as the name of the object, interval, duration, geographical area, etc. Such data can be an event occurring, which is a short description of the sensed phenomenon. The interest is disseminated by a sink through the network, hop-by-hop, to its neighbors. Each node receiving the interest can do caching of it for later use. Each sensor node that receives the interest setup is a gradient toward the sensor nodes from which it receives the interest. This process continues until the gradients are set up from the sources back to the sink sensor node. A gradient is a reply link that specifies an attribute value, a direction, and an expiration time derived from the field of the received interest. After the initial interest and gradients, a multipath is established so that one of the paths is selected by reinforcement. The sink node resends the original interest messages through the selected path with a smaller interval, so it reinforces the source sensor node on that path to send data more frequently. The authors claimed that the DD achieves its full potential, and careful attention has to be paid to the design of sensor radio MAC layers. Many principles are used in the design of WSN protocols, such as self-detection of link quality, lower power availability, and in-network processing. An improved distributed algorithm that can dynamically select optimal paths, called expected transmission count-delay (Expected Transmission Delay GrEedy [EDGE]), was presented in Reference 138. The proposed algorithm was based on DD for computing an aggregate link layer metric, namely EXT, for path selection to increase end-to-end throughput in transmission for supporting demanding applications.

Many restrictions in supporting real-time multimedia streaming applications in WSNs exist. Multiple description coding (MDC) was considered in Reference 139 as a recent advance in multimedia source coding and inexpensive hardware, such as complementary metal–oxide semiconductor (CMOS) cameras and microphones have made a multimedia transmission over a multipath routing wireless sensor protocol possible.

The authors in Reference 15 focused on energy efficiency by proposing the **Energy-Aware Directed Diffusion (EADD)** protocol. The protocol prevents the duplication of forwarding messages (DFM) that causes a reduction of the energy consumption of sensor nodes and causes an imbalance of AE distribution in the network. To avoid both DFM and AE, EADD selects the path to forward a packet according to the average AE value of each sensor node that is distributed in the overall sensor node network. Therefore, if a sensor node has a surplus energy, the sensor node can get a faster response time. Then, the EADD begins to run after receiving the interest message. Through the original protocol that has caused the unbalanced energy distribution, the EADD achieves faster gradient setup and reinforcement.

2.7.2.2 Discussion

Data are routed between multiple sources and sinks in multi-hop architecture in the sensor network. The sink connects directly to the task manager via satellite or internet gateway. The design of multihop architecture is different from traditional ad hoc routing and is influenced by many factors, such as fault-tolerance, scalability, network topology, hardware constraints, power consumption, and the transmission media. Consequently, the design of routing protocols that work in the network layer of the sensor network is usually achieved according to many principles, such as sensor networks that are mostly data-centric, the functionality of data aggregation, the attribute-based addressing and location awareness, and energy efficiency.

Generally, the advantages of data-centric routing are based on provisions of the nodes from individual addresses, which increase energy efficiency, aggregation, and caching data, and on minimizing the delay during data transmission through multiple paths. Data-centric routing uses attribute-based naming to carry out queries by using the attributes of the phenomenon [140].

This attribute basis creates broadcasting, multicasting and anycast for sensor network techniques. Unlike traditional end-to-end QoS routing, data-centric routing tries to find multipaths from a source or multiple sources and then applies the aggregation function at the sink node. Thus, the sink diffuses some interest messages to describe a task that must be performed by the network in multi-hop fashion. This process repeats until the sink is reached. Unfortunately, this technique does not take into account constraints such as energy, unreliability of low-power wireless links, etc., which are imposed by WMSNs. Therefore, data-centric routing uses the data aggregation to overcome the implosion problem that is caused from the duplication of packets that are sent to the same node.

In addition to the implosion problem, the data-centric routing used another technique known as gossiping to overcome the overlapping problem by the sensor node to randomly select one of its neighbors and to send the packets to it. This procedure repeats until all nodes receive these packets. Although this technique tackles

Table 2.8   Comparison of Multipath Broadcast Routing/ Data-Centric Routing Protocols for WSNs

Routing Protocol	Scalability	Energy Efficiency	Reliability	QoS
DD [37]	Low	High	Moderate	†
EADD [15]	Low	High	High	†
EDGE [138]	Low	Moderate	Moderate	†

the implosion problem by having just one copy of a message at any node, there is a significant delay to propagate the packet to reach all sensor nodes in a network.

The major disadvantage of data-centric routing is that it is usually based on a flat-topology structure that causes many problems, such as scalability, increased traffic congestion among the nodes much closer to the sink (known as the broadcast storm), and an increase in overhead complexity. Therefore, clustering was introduced to subdivide the broadcast area into smaller cluster areas. Another disadvantage is in the distributed aggregation mechanisms, which are more applicable for query application models in WSNs. At the same time, directed diffusion is not a perfect choice for time-driven (continuous) models or even for environmental monitoring, an important application in WSNs. Several routing protocols in WSNs are data-centric. However, it is difficult to enumerate all of them in this survey. Therefore, Tables 2.8 and 2.9 depict summaries and compare the aforementioned multipath-broadcast forwarding-routing protocols that are most related to the data-centric problem based on the QoS constraints, data delivery, network architecture, and other parameters, where the cross sign indicates that QoS has been considered by the listed routing protocol.

2.7.3 On-Demand Fashion

Four additional categories including reactive, proactive, hybrid, and geographic routing approaches are strongly dependent on how the source and the route to

Table 2.9   Summary of Multipath Mechanism for Alternative Multipath Broadcast Flooding/Data-Centric Routing Protocols for WSNs

Routing Protocol	Path Disjointedness	Number of Paths	Path Selection	Traffic Distribution
DD [37]	Partially disjoint	Not limited	Sink	Not applicable
EADD [15]	Node-disjoint	Not limited	Sink	Not applicable
EDGE [138]	Partially disjoint	Not limited	Sink	Not applicable

the final destination are discovered [141]. In the proactive method, all paths are computed before they are really needed. This calculation is required for periodic information exchange to update the cause generating a large number of control messages. For this reason, the proactive method is not suitable for a mobile ad hoc network. Somehow, it is preferable for WSNs when the sensor nodes are static [99]. In the reactive methods, the paths are based on discover on demand.

The authors in Reference 99 surveyed multipath routing protocols for mobile ad hoc networks through classification that depends on many objectives of a multipath routing protocol, such as high end-to-end delay, unreliable data packet transfer, energy inefficiency, high overhead, and scalability.

This section limits discussion only to proactive and reactive multipath protocols that are preferable for WSN QoS-routing protocols and are mostly cited to originate from dynamic source routing (DSR) and the ad hoc on-demand distance vector (AODV). Although there are similarities between the MANETs and WSNs, most multipath routing protocols are designed without explicitly considering the QoS parameters of the paths that they generate. That is, not only is the selected path from source to destination optimized, but the end-to-end QoS requirements are also guaranteed, often in terms of reliability, overhead reduction, power consumptions, delay, and throughput.

2.7.3.1 DSR Principal Protocols

One of the many routing wireless ad hoc characteristics is in allowing an ad hoc network to establish on-the-fly building of unconstrained connectivity and fault tolerance. The authors in Reference 142 focused on designing a QoS metric and end-to-end reliability through distributed on-demand multipath dynamic source routing (MP-DSR).

Robust multipath routing for dynamic topology in WSNs is discussed in Reference 38: *Such characteristics allow an ad hoc net-work to be established on-the-fly with built-in fault tolerance and unconstrained connectivity.* The authors provided efficient fault tolerance using robust multipath dynamic source routing (RMDSR) for WSNs to increase the recovering operation from route failure in the dynamic network domain.

The authors in Reference 143 introduced a routing algorithm for WSNs. This work was performed as a part of the EYES project (IST-2001-34734) on self-organizing and collaborative energy-efficient sensor networks. The algorithm addresses the convergence of distributed information processing, wireless communication, mobile computing, and maintaining the amount of overhead traffic at a low value.

The authors in Reference 144 introduced a new metric called network survivability for the reactive-routing protocol performance. This metric implies that the protocol should ensure that connectivity in a network is maintained for as long as possible and that the energy health of the entire network should be of the same order. It is necessary to take care of network balancing, connectivity, and energy level. Leaving the network might cause a wide disparity in the energy levels of the nodes and eventually may lead to network partition.

2.7.3.2 AODV Principal Protocols

An ad hoc protocol builds on the idea of the distance-vector table algorithm with little modification in either route discovery or route maintenance by minimizing the number of required broadcasts using routes on an on-demand basis. A few of the studies whose scopes are in the multipath routing approach are presented below.

The author in Reference 39 proposed a tiny optimal node-disjoint multipath routing (TinyONDMR) protocol that is suitable for transmission of larger multimedia with better performance, reliability, and efficiency in a WSN. The routing protocol optimizes the node-disjoint by finding multi-nodes disjoint between the sources and the sink that lead to an optimal load balancing. The TinyONDMR, inspired by the modification of split multipath routing and the DSR, achieves low-routing overhead during route discovery.

The concept of route coupling, which means transmission interference between the multipath routing or even the single-path routing between two node-disjoints, was discussed in Reference 145. This interference leads to degradation of network performance, limiting the QoS parameters, etc. This phenomenon occurs when two paths are located within each other during the beginning of transmissions. Therefore, to save the power of the sensor node, traffic routed along the multipath does not interfere with other traffic. The authors have also proposed optimized cross-layers between the MAC layer and network routing for a new energy-efficient multipath approach [146], which was inspired by the ad hoc on-demand multipath distance vector (AOMDV) and [145].

A resilient multipath routing protocol (RMRP) was presented in Reference 40. RMRP uses a scheme for maintaining broken links by reducing the flooding range of the control messages.

2.7.3.3 Proactive Routing

In the proactive routing protocols (or table-driven routing protocols), each node attempts to maintain consistency before establishing paths with every other node in the network with up-to-date routing information [147]. The main advantage of this approach is that the routes are available whenever they are needed, and no delays are incurred in searching for routes in on-demand routing protocols. The proactive approach responds to changes in the network topology or another metric, such as bandwidth and interference, by propagating updates throughout the network to maintain a consistent network [147]. The main disadvantages are that the overheads involved in building and maintaining potentially very large routing tables and the stale information in these tables may lead to routing errors.

The architecture of transmission control protocol/Internet protocol (TCP/IP) in traditional wire networks is not suitable for WSNs for many reasons, such as limitation on storage against a small size, low-cost, power consumption, reliability, etc. Thus, involving sensor network in monitoring and controlling activities faces many challenges, which are addressed by a new protocol SHRP [16]. This hierarchical and proactive routing protocol models the performance of the energy of the sensor and reliability metrics to maintain the WSN topology. This protocol monitors the battery availability and link quality. The mechanism of monitoring is based on two measured metrics, link quality indicator (LQI) and received signal strength indicator (RSSI) that are offered by the data link layer.

2.7.4 Discussion

The differences between ad hoc and sensor networks should be considered as the primary concern in designing an efficient routing algorithm based on an ad hoc manner for a WSN. Usually, ad hoc wireless networks possess high network dynamic topology specifications owing to mobility and failures. The design of distributed efficient algorithms to dynamically update the routing structures is considered a critical issue in the implementation of WSNs. For example, the proactive routing protocol is periodically required for each node to maintain and update one or more tables to store routing information during changes in network topology, which leads to extraneous energy consumption and wastage in bandwidth because of the higher control packet overhead. Reactive protocols are on-demand routing protocols. This means that they do not periodically update and hence require less routing information for each node. Therefore, the provision of energy and bandwidth is better than proactive routing protocols during inactivity [148].

In real-time applications, proactive routing protocols are more appropriate because they do not require a latency in route discovery, unlike reactive routing protocols [42]. However, there is a latency for discovering the route that is called acquisition delay that may not be suitable for real-time applications [59]. Traffic loading in WSNs could be either broadcast or convergecast, that is, high loads or bursts of traffic. Usually, the broadcast traffic is widely used in various wide network queries and updates. That is the reason why broadcast traffic is often observed when multi-sources have detected the same event and transmit the data packet to a node that does require data aggregation [149]. Therefore, reactive routing protocols are mainly optimized for light traffic loads if the routing information changes frequently and if route discoveries are not needed for those routes changes [150]. Otherwise, reactive routing protocols may result in a large volume of messaging overhead [141].

Unlike the reactive route protocols, the availability of changing route information is considered a key advantage of the proactive routing protocols. The protocols become faster in routing decisions, more power-efficient, packet delivery ratio, and

Table 2.10 Comparison of Multipath Mechanism for Alternative Multipath Broadcast Flooding/On-Demand Fashion Routing Protocol for WSNs

Routing Protocol	Scalability	Energy Efficiency	Reliability	Network Topology	Latency	Network Dynamic	Localization
MP-DSR [142]	Low	Low	High	Flat	Low	†	†
RMDSR [38]	Low	Low	High	Flat	Moderate	†	†
TinyONDMR [39]	Low	Low	High	Flat	Low	†	†
RMRP [40]	Low	Low	High	Flat	Low	†	†
SHRP [16]	Low	High	High	Hierarcy	Low	†	†

(consequently) latency in route discovery for heavy traffic loads compared to reactive protocols.

Another important advantage of proactive routing protocols is that periodic routing updates keep the routing for each node up-to-date. This advantage reduces the cost of higher signaling traffic than that required by reactive routing protocols. Furthermore, the sensor nodes spend more energy because of their periodic update messages. However, there are variations for other operations such as route reconfiguration after failure that depends on the mechanism that the routing protocols use to perform these operations.

For reactive routing protocols, certain QoS parameters do not ensure construction of an optimal route from the source to the sink with guaranteed delivery packets within the specified time. Tables 2.10 and 2.11 depict a summary and comparison of the aforementioned multipath-broadcasting forwarding-routing protocol for nodes in a genetic and ad hoc manner on QoS constraints, data delivery, network architecture, and other parameters, where the cross sign indicates that the parameter has been considered by the listed routing protocol.

Table 2.11 Summary of Multipath Mechanism for Alternative Multipath Broadcast Flooding/On-Demand Fashion Routing Protocol for WSNs

Routing Protocol	Path Disjointedness	Number of Paths	Path Selection	Traffic Distribution
MP-DSR [142]	k-disjoint path	Not limited	Sink	Not applicable
RMDSR [38]	Partially disjoint	Not limited	Sink	Not applicable
TinyONDMR [39]	Node-disjoint	Not limited	Sink	Not applicable
RMRP [40]	Node-disjoint	Not limited	Sink	Not applicable
SHRP [16]	Node-disjoint	Not limited	Sink	Not applicable

2.8 Simulation Comparisons

This section provides a comprehensive analysis of major multipath routing approaches for WMSNs. A multipath routing mechanism is considered as an efficient approach to improve network capacity and resource utilizations under heavy traffic conditions [151]. Therefore, there is certainly an urgent need to analyze the significance as well as the performance analysis in order to identify the challenges pertaining to the design of multipath routing protocols for WMSNs. To demonstrate the effect of selection for a multipath routing mechanism, we have conducted some simulations of major multipath routing approaches, namely Re-InForM [28], MCMP [33], ECMP [34], and PMR [35] in terms of several performance parameters. These parameters include the power consumption, the average end-to-end delay, and the successfully received packet ratio. The success probability of the transmission channel is chosen from 0.1 to 0.8, which implies a good state for the link quality [35]. We used the LINGO optimization module [152] considering a linear WSN with an area (1000 m * 1000 m) composed of *n* sensor nodes. The source sensor node and the sink are placed inside the wireless sensor area; each sensor node has a connectivity that is associated with two positive QoS constraints. We evaluated the object-tracking scenario by studying the behavior of cars passing along the highway in a one-dimensional sensor network topology. The parameters used in the proposed model are presented in Table 2.12.

Table 2.12   Definition of Parameters in Simulation Scenarios

Parameter	Value
The overhead energy due to the sensing, receiving, and processing	50 nJ/bit
The loss coefficient related to *p* bit transmission propagated over single-path model	10 pJ/bitm^2
The loss coefficient related to *p* bit transmission propagated over multipath model	0.0013 pJ/bitm^2
Topology structure	Urban highway, sensor node distributed uniformly
Total number of sensor nodes	50 sensor nodes
Message payload	64 bytes
Data length *p*	2000 bits
Transmission range	12.00 m
Total distance	150 m

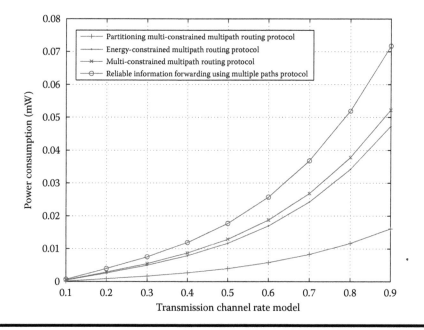

Figure 2.18   Power consumption comparison.

Results in Figure 2.18 show the total power consumption in the network with four multipath routing algorithms. The PMR model performs better than the ECMP and MCMP, due to the fact that both the ECMP and MCMP are deterministic zero-one problems with linear objective functions and constraints. The ECMP model uses a smaller group of longer multipaths than the MCMP model. The MCMP model consumes more energy than the ECMP model, which lowers the average end-to-end delay as seen in Figure 2.19. The similarity of the performances of the ECMP and the MCMP justifies that the trade-off between power consumption and average end-to-end delay is affected by the number of selected multipaths. This finding reveals that the ECMP model uses a smaller or longer multipath, resulting in lower power consumption with a higher average end-to-end delay than the MCMP model. The PMR model outperforms the other models since it uses preferred selected multipath with an optimal hop number. This result reveals that PMR uses the multipath with the optimal hop number. Therefore, the PMR model is more likely to lead to a lower average end-to-end delay than the other two models. All multipath routing approaches perform equally in terms of the PRR for the reliability of transmission channels as seen in Figure 2.20. To have a close view of the PRR, the curves of all models display the PRR on a log scale. Among all the models, the PMR protocol achieves the best PRR of approximately 96%. The ECMP, MCMP, and Re-InForM models achieve PRR values of 94%, 92%, and 88%, respectively. The packets reception ratio usually increases logarithmically as

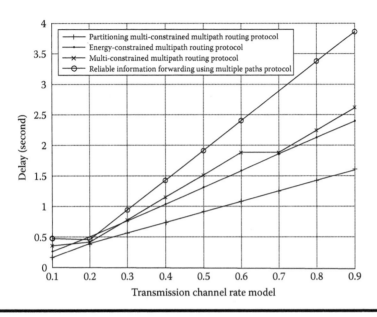

Figure 2.19 Average delivery end-to-end delay comparison.

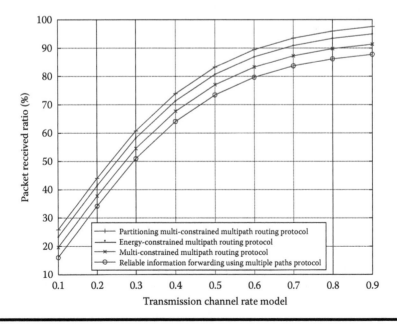

Figure 2.20   Comparison of received packet ratio.

the reliability of the channel transmission rate increases because of the confirmation of more packets that are successfully delivered to the sink, with a small expiration ratio for the lost packets.

A summary of the comparative performance analysis among major multipath routing protocols is listed in Table 2.13. These simulation comparisons address the problem of providing load balancing efficiency, high bandwidth utilization efficiency, and packet order preservation in WMSNs. In particular, we are interested in providing the reader with an understanding of reliable real-time applications, by focusing on the design of a multipath routing protocol for WMSNs that operates efficiently under heavy traffic situations. We describe performance evaluations in terms of the number of paths allowed for transmission toward the sink and not just the best path. This should be achieved without imposing an excessive overhead control problem in maintaining these paths to indicate the performance of the multipath routing protocol. Therefore, the absolute performance is interpreted under the degree for the state of complexity for discovering, selecting, and maintaining multiple paths. The state of complexity is categorized as high, mod., and low, which can be interpreted as follows. High means that the performance of the routing protocol degrades dramatically due to a problem in routing operations. Mod. indicates the problem may not occur. Low means that the problem does not cause any significant impact.

2.9 Open Research Problems

In spite of the numerous research activities and the remarkable progress in recent years, multipath routing protocols in WMSNs still harbor many open issues that are still waiting to be resolved. The main goals of a multipath routing protocol include carrying out data communication while maximizing data delivery, extending the network lifetime, minimizing energy expenditure, and preventing connectivity degradation. These goals can be achieved by employing data aggregation, energy management, and efficient control of path selection techniques. Additionally, there are restrictions on the nodes and multimedia content that impose additional challenges. Therefore, there are extensive research issues that should further be explored in the field of multipath routing protocols for WMSNs. In this section, a broad range of research issues are outlined for future investigation.

2.9.1 Data Sensing and Delivery Model

The data sensing and delivery model affect the performance of the multipath routing protocols. Especially with regard to energy consumption and path selection, that is, if a node is continually capturing a video content and sending it to the sink, the node consumes more energy. Therefore, it reduces its own lifetime and, consequently, that of the whole system as well. Depending on the type of the multipath

Table 2.13 Simulation Comparison of Multipath Routing Protocol for WMSNs

Protocol	Approach Used	Simulator Type	Performance Metrics	Comparison Results	State Complexity
SAR [17]	Load balancing and support for multiple priorities.	Parsec	QoS and node lifetime	SAR has achieve a better performance than the minimum metric algorithm.	Mod.
SPEED [18,19][a]	Back-pressure rerouting to control congestion.	Theoretical analysis, GloMoSim, and real implementation on Berkeley motes	Delay and packet loss ratio	SPEED performs better in terms of end-to-end delay ratio and packet lost ratio.	Mod.
REAR [21]	Establish multipath routing for reducing energy consumption.	Not specified	Delay and network lifetime	REAR prolongs the network life cycle better than SAR.	High
RPAR [10]	To achieve application specified communication delay at low energy cost by dynamically adjusting transmission power and routing decisions.	MATLAB-based network simulator called Prowler	Energy and delay	RPAR employs a novel neighborhood management mechanism that is more efficient than the periodic beacon scheme adopted by SPEED and MMSPEED	High
MMSPEED [27][a]	Supports multipath routing and multiple network-wide packet delivery speeds.	J-SIM network simulator	Delay and reliability	MMSPEED could use its redundant path selection scheme for load balancing, which is not only for reliability enhancement but also to improve the overall network lifetime.	Mod.

(Continued)

Table 2.13 (*Continued*) Simulation Comparison of Multipath Routing Protocol for WMSNs

Protocol	Approach Used	Simulator Type	Performance Metrics	Comparison Results	State Complexity
Re-InForM [28]	Propose approaches based on multipath routing to provide reliability.	Not specified	Energy and reliability	Re-InForM achieved packet delivery probability with a 40% and 70% reliability target compared with two other schemes flooding broadcasting and sending only a single packet over a single path.	High
NC-RMR [29]	Presents network coding-based reliable disjoint and braided multipath routing for sensor networks in which improving reliability of multipath routing with network coding is the primary aim.	OMNet++	Energy and reliability	The protocol improves data transmission reliability and saves energy consumption. The performance of reliability and normalized energy consumption is studied by analysis and simulation methods.	High
DMRF [31]	Demonstrate a dynamical jumping real-time fault tolerant routing protocol.	Prowler simulation platform	Delay and packet received ratio	DMRF can achieve the successful transmission ratio up to 92% more often than SPEED and MMSPEED	Low
MCMP [33]	Utilize the multiple paths between the source and sink pairs for QoS provisioning.	Theoretical analysis and PARSEC	Delay and packet received ratio	MCMP outperforms single-path routing remarkably and approaches approximately 95% of that of Mod. routing and braided multipath routing	Low

(Continued)

Table 2.13 (*Continued*) Simulation Comparison of Multipath Routing Protocol for WMSNs

Protocol	Approach Used	Simulator Type	Performance Metrics	Comparison Results	State Complexity
ECMP [34]	Models QoS routing in WSNs to achieve energy efficiency.	Theoretical analysis	Delay and packet received ratio	The MCMP algorithm shares its traffic on more paths than the ECMP algorithm; however, the ECMP algorithm can achieve more energy savings than the MCMP model	Low
PMR [35]	Partitioning multipath routing to achieve QoS parameters.	Theoretical analysis	Energy consumption, delay, and packet received ratio	PMR provides sufficient information about the links between sensor nodes to determine the optimal path and to select the intermediate sensor node for routing the packets from source to sink. The results demonstrate that PMR improves the PRR from the source to the sink, increasing the lifetime and minimizing the end-to-end delay.	Low

(Continued)

Table 2.13 (*Continued*) Simulation Comparison of Multipath Routing Protocol for WMSNs

Protocol	Approach Used	Simulator Type	Performance Metrics	Comparison Results	State Complexity
DD [37]	To describe and illustrate one instantiation of the paradigm for sensor query dissemination and processing. To empirically adapt to a small subset of network paths and achieve significant energy savings when intermediate nodes aggregate responses to queries.	Theoretical analysis and NS-2 simulator	Energy, average delay, and distinct-event delivery ratio	DD has noticeably better energy efficiency than omniscient multicast; moreover, DD reduces data transmission delay caused by path failure by decreasing the frequency of path rediscovery [30].	Mod.

a Both SPEED and MMSPEED have a common deficiency: the energy consumption metric has not been taken into account.

routing protocol, continuous data transmission causes a path selection. A similar kind of behavior occurs with scalar data transmissions.

2.9.2 Node Deployment

The distance between two sensor nodes affects link quality. In the case of a network in which some nodes have a small number of neighborhoods, this node rapidly exhausts its battery. Additionally, for a nonuniform distribution, an optimal position of the sources and the sink are necessary to allow connectivity that can enable an energy-efficient network operation. Because multimedia sensor nodes are sensitive to direction of acquisition and have limited coverage, a multimedia sensor node must be deployed in the best place to optimize the coverage and avoid obstacles. Moreover, several multimedia sensor nodes are equipped with a camera, and there is a need to study the best place to deploy them to ensure an optimal coverage. The other nodes can be deployed in a random way, although with a uniform distribution [153].

2.9.3 Node Capabilities

Several sensor nodes can have different roles or capabilities according to the application, which is related to the capacity of the nodes in terms of computation, communication, power, and multimedia support. Most of the applications use homogeneous nodes, which have equal capabilities or are produced by the same manufacturer. However, in some applications the network is considered to be heterogeneous because some/all of the nodes have different capabilities or roles. In this context, they are able to perform special functions, such as sensing, aggregation, or the retrieval of multimedia content. Therefore, in the case of a heterogeneous network, the node that is able to perform many tasks, for example, sensing, aggregation, or retrieval of multimedia content, is likely to end up its source of energy in a short period of time. As mentioned previously, several of the nodes are constrained in terms of their processing capability, which makes it difficult to carry out many tasks. To overcome challenges arising from multimedia content and the restrictions of the sensor node, the use of a heterogeneous network can be an alternative solution.

2.9.4 Link Quality Estimators (LQEs)

Wireless links in WMSNs are unreliable and unpredictable. This is mainly due to the fact that the nodes use low-power radios, which are sensitive to noise, interference, and multipath distortion. Therefore, it is necessary to quantify a metric for communications between neighborhoods. This metric is obtained through a link quality estimator (LQE). A path is considered to be good when a link has the highest value. Several multipath routing protocols rely on LQEs as a mechanism for selecting the most stable routes, and the accuracy of the selected LQE has an impact on their performance. Thus, LQE is a fundamental building block in the design of routing protocols for WMSNs.

2.9.5 Mobility

Mobility is one of the key challenges in wireless communications, since the problem of routing messages arises throughout the network. Because paths are continually changing, this leads to an important issue regarding optimization, as well as in regard to achieving improvements in energy saving and bandwidth. Moreover, in cases of dense WMSNs, this is one of the main challenges, since it would be very difficult for a sensor node to keep track of all the neighboring nodes because of the fact that every node may have a large number of mobile neighbors that dynamically change the network topology. Therefore, more research into an efficient mobility management scheme would enhance the performance of the WMSNs in terms of energy saving and bandwidth utilization.

2.9.6 Scalability

Scalability is one of the main design attributes of WMSNs and must be encompassed by the routing protocols. However, the routing protocol should be scalable enough to enable it to work with a large number of sensor nodes and to continually ensure the correct behavior of the application. Furthermore, it must be adapted to scalability changes in a transparent way, that is, without requiring intervention of the user. To achieve efficient data processing, aggregation, storage, and querying in WMSNs, the scalability must be taken into account, especially when this involves a large amount of multimedia data. Therefore, network designers must employ a hierarchical architecture that offers considerable advantages with respect to a flat architecture in terms of scalability, lower costs, better coverage, higher functionality, and greater reliability.

2.9.7 Multimedia Content

Multimedia content produces a large amount of data. The nodes in a WMSN have the capability to capture and transmit multimedia content, which can either be a snapshot or streaming content. Actually, a snapshot contains data from an event that was triggered in a short time period. Streaming multimedia content is generated over a longer time period. Thus, transmitting video as a snapshot or streaming requires a high data rate over the WMSN link, which is extremely difficult. Because of the limitations of the sensor nodes, process video coding/compression with a low complexity produces a bandwidth with a low output, which can tolerate loss, and consumes as little power as possible. Furthermore, the predictive schemes for video coding techniques are not suitable for WMSNs, as they require complex encoders and powerful processing algorithms, and entail high energy consumption. However, a multipath routing mechanism can be adopted to increase the bandwidth. The multimedia content will only be sent when events occur to reduce the amount of data.

2.9.8 Energy Efficiency Considerations

Multimedia traffic represents a large amount of data which has to be delivered over the network. Hence, the majority of routing protocols in WMSNs consider energy efficiency as the main objective and assume data traffic with unconstrained delivery requirements [51]. Because there is a risk that the node routing multimedia data might quickly drain its energy, the main challenge is to achieve reliability in data delivery packets, with a minimal consumption of energy, while increasing the lifetime of the network.

2.9.9 Multi-Constrained QoS Guarantee

QoS specification can be used to provide various priorities to guarantee a certain level of performance to a data flow in accordance with requests from the application. However, a QoS guarantee is considered necessary, especially when the network capacity is limited. For example, real-time streaming multimedia applications require a fixed bit rate and are delay sensitive [51]. Therefore, routing algorithms offering a QoS guarantee for multimedia traffic should be flexible to support different application-specific QoS requirements (such as energy efficiency, end-to-end delay, reliability, delay jitter, and bandwidth consumption) in the heterogeneous traffic environment [54].

2.9.10 Cognitive Radio (CR)

CR is based on dynamic spectrum access (DSA), which has been proved to yield a promising approach for communications [154]. CR has positive impacts on power consumption levels and network lifetime [155]. Capabilities such as sensing the spectrum, determining the vacant band, and improving the overall utilization may be exploited by WMSN to employ spectrum allocation and processing resource constraints of low-end of WMSN sensor nodes [156]. In other words, WMSNs are enriched with these additional capabilities of CR and may lead to an evolution of cognitive radio sensor networks (CRSNs) according to applications such as defense, utility metering, and home automation [157,158]. Each sensor node may be equipped with DSA to detect an event-driven communication pattern that generates trusty traffic that needs to efficiently utilize the channel [158]. Due to CRSNs having large bandwidth requirements, WMNSs can be integrated with low-cost hardware and enjoy the benefits of CR to satisfy the network and application requirements [158]. CRSNs face certain challenges that must be investigated for specific design considerations, existing approaches, and open research issues of each different layer in the protocol stack of the sensor network [159]. For example, the goal of most existing CR routing protocols is to provide joint spectrum and routing decisions, but these do not consider the inherent resource constraints of CRSNs [160]. However, the network designer should focus on developing new metrics, such as channel access delay, interference, and bandwidth, and mechanisms that consider resource constraints and dense deployment in order to design new energy-efficient routing protocols, whether it be through a single-path or multipath approach [160].

2.10 Conclusion

The emergence of multipath routing techniques in supporting a wide range of real-time multimedia applications has attracted the attention of several researchers worldwide. The common objective is to develop new protocols that guarantee network performance by providing QoS. In this survey, we have introduced a new taxonomy on multipath-routing protocols. The structure of the taxonomy is based on the employed path utilization methods that are used by multipath routing to improve the capacity and resource utilization of the network. We have highlighted the advantages and disadvantages of several routing mechanisms and algorithms. The development of efficient multipath routing protocols with multi-constraints for WMSNs is an open research area that is promising for future research.

Acknowledgments

The authors would like to sincerely acknowledge the anonymous reviewers for their very constructive and helpful comments and suggestions which have ultimately resulted in significant improvement of the content and quality of the final manuscript.

References

1. S. Prabhavat, H. Nishiyama, N. Ansari, and N. Kato, On load distribution over multipath networks, *IEEE Communications Surveys & Tutorials*, vol. 14, pp. 662–680, 2012.
2. I.F. Akyildiz et al., A survey on wireless multimedia sensor networks, *Computer Networks*, vol. 51, pp. 921–960, 2007.
3. A. Alanazi and K. Elleithy, An optimized hidden node detection paradigm for improving the coverage and network efficiency in wireless multimedia sensor networks, *Sensors*, vol. 16, no. 9, pp. 1438–1454, 2016.
4. E. Al Nuaimi, H. Al Neyadi, N. Mohamed, and J. Al-Jaroodi, Applications of big data to smart cities, *Journal of Internet Services and Applications*, vol. 6, pp. 1–15, 2015.

5. B. Rashid and M. H. Rehmani, Applications of wireless sensor networks for urban areas: A survey, *Journal of Network and Computer Applications*, vol. 60, pp. 192–219, 2016.

6. M. A. Alsheikh, D. T. Hoang, D. Niyato, H.-P. Tan, and S. Lin, Markov decision processes with applications in wireless sensor networks: A survey, *IEEE Communications Surveys & Tutorials*, vol. 17, pp. 1239–1267, 2015.

7. M. Abazeed et al., Routing Protocols for wireless multimedia sensor network: A survey, *Journal of Sensors*, vol. 2013, pp. 1–11, 2013.

8. A. Ali et al., Real-time routing in wireless sensor networks, in *Proceedings of the 28th International Conference on Distributed Computing Systems Workshops (ICDCS)*, pp. 114–119, 2008.

9. J. Zhu et al., Tasks allocation for real-time applications in heterogeneous sensor networks for energy minimization, in *Proceedings of the Eighth ACIS International Conference on Software Engineering, Artificial Intelligence, Networking, and Parallel/Distributed Computing*, vol. 02, pp. 20–25, 2007.

10. O. Chipara et al., Real-time power-aware routing in sensor networks, in *14th IEEE International Workshop on Quality of Service, IWQoS*, pp. 83–92, 2006.

11. M. S. Al-Fares and S. Zhili, Self-organizing routing protocol to achieve QoS in wireless sensor network for forest fire monitoring, in *IEEE 9th Malaysia International Conference on Communications (MICC)*, pp. 211–216, 2009.

12. J. Ben-Othman and B. Yahya, Energy efficient and QoS based routing protocol for wireless sensor networks, *Journal of Parallel and Distributed Computing*, vol. 70, pp. 849–857, 2010.

13. Z. Yuanyuan et al., A real-time and robust routing protocol for building fire emergency applications using wireless sensor networks, in *8th IEEE International Conference on Pervasive Computing and Communications Workshops (PERCOM Workshops)*, pp. 358–363, 2010.

14. Y. Hu, Y. Ding, and K. Hao, An immune cooperative particle swarm optimization algorithm for fault-tolerant routing optimization in heterogeneous wireless sensor networks, *Journal of Mathematical Problems in Engineering*, vol. 2012, p. 19, 2012.

15. C. Jisul and K. Keecheon, EADD: Energy aware directed diffusion for wireless sensor networks, in *International Symposium on Parallel and Distributed Processing with Applications (ISPA)*, pp. 779–783, 2008.

16. C. J. Barenco Abbas et al., SHRP: A new routing protocol to wireless sensor networks, in *Advances in Computer Science and Engineering*, vol. 6, H. Sarbazi-Azad et al., Eds., Springer: Berlin, Germany pp. 138–146, 2009.

17. K. Sohrabi et al., Protocols for self-organization of a wireless sensor network, *IEEE Personal Communications*, vol. 7, pp. 16–27, 2000.

18. H. Tian et al., SPEED: A stateless protocol for real-time communication in sensor networks, in *Proceedings 23rd International Conference on Distributed Computing Systems*, pp. 46–55, 2003.

19. T. He et al., A spatiotemporal communication protocol for wireless sensor networks, *IEEE Transactions on Parallel and Distributed Systems*, vol. 16, pp. 995–1006, 2005.

20. M. S. Kordafshari, A. Pourkabirian, K. Faez, and A. M. Rahimabadi, Energy-efficient SPEED routing protocol for wireless sensor networks, in *Fifth Advanced International Conference on Telecommunications (AICT)*, pp. 267–271, 2009.

21. L. Yao et al., A real-time and energy aware QoS routing protocol for multimedia wireless sensor networks, *Proceedings of the 7th World Congress on Intelligent Control and Automation (WCICA)*, pp. 3321–3326, 2008.

22. N. Saxena et al., QuESt: A QoS-based energy efficient sensor routing protocol, *Wireless Communications and Mobile Computing*, vol. 9, pp. 417–426, 2009.

23. Y.Y. Li, C.S. Chen, Y.Q. Song, Z. Wang, and Y. Sun, Enhancing real-time delivery in wireless sensor networks with two-hop information, *IEEE Transactions on Industrial Informatics*, vol. 5, pp. 113–122, 2009.

24. J. Park et al., Disjointed multipath routing for real-time data in wireless multimedia sensor networks, *International Journal of Distributed Sensor Networks*, vol. 2014, p. 8, 2014.

25. J. Agrakhed et al., Adaptive multi constraint multipath routing protocol in wireless multimedia sensor network, in *Proceedings of the International Conference on Computing Sciences (ICCS)*, pp. 326–331, 2012.

26. I. Nikseresht et al., Interference-aware multipath routing for video delivery in wireless multimedia sensor networks, in *Proceedings of the 32nd International Conference on Distributed Computing Systems Workshops (ICDCSW)*, pp. 216–221, 2012.

27. E. Felemban et al., MMSPEED: Multipath Multi-SPEED protocol for QoS guarantee of reliability and. Timeliness in wireless sensor networks, *Mobile Computing, IEEE Transactions on*, vol. 5, pp. 738–754, 2006.

28. B. Deb et al., ReInForM: Reliable information forwarding using multiple paths in sensor networks, in *2003. LCN '03. Proceedings. 28th Annual IEEE International Conference on in Local Computer Networks*, pp. 406–415, 2003.

29. Y. Yang et al., Network coding based reliable disjoint and braided multipath routing for sensor networks, *Journal of Network and Computer Applications*, vol. 33, pp. 422–432, 2010.

30. U. Mahadevaswamy and M. Shanmukhaswamy, An energy efficient reliable multipath routing protocol for data gathering in wireless sensor networks, *International Journal of Computer Science and Information Security (IJCSIS)*, vol. 8, no. 2, pp. 59–64, 2010.

31. G. Wu et al., Dynamical jumping real-time fault-tolerant routing protocol for wireless sensor networks, *Sensors*, vol. 10, pp. 2416–2437, 2010.

32. C. Yunfeng and N. Nasser, Enabling QoS multipath routing protocol for wireless sensor networks, in *ICC '08. IEEE International Conference on Comminications*, pp. 2421–2425, 2008.

33. X. Huang and Y. Fang, Multiconstrained QoS multipath routing in wireless sensor networks, *Wireless Networks*, vol. 14, pp. 465–478, 2008.

34. A.B. Bagula and K.G. Mazandu, Energy constrained multipath routing in wireless sensor networks, in *Ubiquitous Intelligence and Computing*, Springer: Berlin, Germany, pp. 453–467, 2008.

35. M.Z. Hasan and W.T. Chee, Optimized quality of service for real-time wireless sensor networks using a partitioning multipath routing approach, *Journal of Computer Networks and Communications*, vol. 2013, pp. 2090–7141, 2013.

36. T. Houngbadji and S. Pierre, QoSNET: An integrated QoS network for routing protocols in large scale wireless sensor networks, *Computer Communications*, vol. 33, pp. 1334–1342, 2010.

37. C. Intanagonwiwat et al., Directed diffusion for wireless sensor networking, *Networking, IEEE/ACM Transactions on*, vol. 11, pp. 2–16, 2003.

38. P. Huang et al., Robust multipath routing for dynamic topology in wireless sensor networks, *The Journal of China Universities of Posts and Telecommunications*, vol. 14, pp. 1–5, 2007.
39. S.-R. Jung et al., An optimized node-disjoint multipath routing protocol for multimedia data transmission over wireless sensor networks, in *Proceedings of the IEEE International Symposium on Parallel and Distributed Processing with Applications*, vol. 33, no. 11A, pp. 1021–1033, 2008.
40. K.-H. Kim et al., A resilient multipath routing protocol for wireless sensor networks, in *4th International Conference on Networking, Reunion Island, France, Proceedings, Part II*, P. Lorenz and P. Dini, Eds., Springer: Berlin, Germany, pp. 1122–1129, April 17–21, 2005.
41. E. Gelenbea et al., Routing of high-priority packets in wireless sensor networks, in *IEEE Second International Conference on Computer and Network Technology*, vol. 66, pp. 1–9, 2008.
42. J. N. Al-Karaki and A. E. Kamal, Routing techniques in wireless sensor networks: A survey, *IEEE Wireless Communications*, vol. 11, pp. 6–28, 2004.
43. I.T. Almalkawi et al., Wireless multimedia sensor networks: Current trends and future directions, *Sensors*, vol. 10, no. 7, pp. 6662–6717, 2010.
44. D. Niyato, E. Hossain, and A. Fallahi, Sleep and wakeup strategies in solar-powered wireless sensor/mesh networks: Performance analysis and optimization, *IEEE Transactions on Mobile Computing*, vol. 6, pp. 221–236, 2007.
45. A. Sutagundar et al., Energy efficient multipath routing protocol for WMSNs, *International Journal of Computer and Electrical Engineering*, vol. 2, no. 3, pp. 503–510, 2010.
46. M. Radi et al., Multipath routing in wireless sensor networks: Survey and research challenges, *Sensors*, vol. 12, pp. 650–685, 2012.
47. S. Ehsan and B. Hamdaoui, A Survey on energy-efficient routing techniques with QoS assurances for wireless multimedia sensor networks, *IEEE Communications Surveys & Tutorials*, vol. 14, pp. 265–278, 2012.
48. W. Lou et al., Performance optimization using multipath routing in mobile ad hoc and wireless sensor networks, *Combinatorial optimization in communication networks*, vol. 18, M. Cheng et al., Eds., Springer: US, pp. 117–146, 2006.
49. N. Mishra and A. Kumar, Comparative analysis: Energy efficient multipath routing in wireless sensor network, *IJCSMC*, vol. 3, no. 9, pp. 627–632, 2014.
50. P. Gopi, Multipath routing in wireless sensor networks: A survey and analysis, *IOSR Journal of Computer Engineering*, vol. 16, pp. 27–34, 2014.
51. K.C. Vairam.T, Interference aware multi-path routing in wireless sensor networks, *International Journal of Emerging Science and Engineering (TM)*, vol. 1, no. 10, pp. 74–78, 2013.
52. M. Korkalainen, M. Sallinen, N. Karkkainen, and P. Tukeva, Survey of wireless sensor networks simulation tools for demanding applications, in *Proceedings of the 5th International Conference on Networking and Services (ICNS)*, pp. 102–106, Valencia, Spain, April 2009.
53. C. P. Singh et al., A survey of simulation in sensor networks, in *International Conference on in Computational Intelligence for Modelling Control & Automation*, pp. 867–872, 2008.
54. Y. Li, C. S. Chen, Y.-Q. Song, and Z. Wang, Real-time QoS support in wireless sensor networks: A survey, in *7th IFAC Proceedings International Conference on Fieldbuses & Networks in Industrial & Embedded Systems*, vol. 40, pp. 373–380, 2007.

55. J. A. Stankovic et al., Real-time communication and coordination in embedded sensor networks, in *Proceedings of the IEEE*, vol. 91, pp. 1002–1022, 2003.
56. A.-D. Zhan, T.-Y. Xu, G.-H. Chen, B.-L. Ye, and S.-L. Lu, A survey on real-time routing protocols for wireless sensor networks, *Chinese Journal of Computer Science*, vol. 3, pp. 234–238, 2008.
57. P. Hambarde, R. Varma, and S. Jha, The survey of real time operating system: RTOS, in *Proceedings-International Conference on Electronic Systems, Signal Processing, and Computing Technologies (ICESC)*, pp. 34–39, 2014.
58. I. F. Akyildiz et al., A survey on sensor networks, *IEEE Communications Magazine*, vol. 40, pp. 102–114, 2002.
59. K. Akkaya and M. Younis, A survey on routing protocols for wireless sensor networks, *Ad Hoc Networks*, vol. 3, pp. 325–349, 2005.
60. J. Lessmann, P. Janacik, L. Lachev, and D. Orfanus, Comparative study of wireless network simulators, in *ICN Seventh International Conference on Networking*, pp. 517–523, 2008.
61. Z. Jin et al., A Survey on position-based routing algorithms in wireless sensor networks, *Algorithms*, vol. 2, pp. 158–182, 2009.
62. C. Li et al., A survey on routing protocols for large-scale wireless sensor networks, *Sensors*, vol. 11, pp. 3498–3526, 2011.
63. N. A. Pantazis, S. A. Nikolidakis, and D. D. Vergados, Energy-efficient routing protocols in wireless sensor networks: A survey, *IEEE Communications Surveys & Tutorials*, vol. 15, pp. 551–591, 2013.
64. Z. Hamid and F. B. Hussain, QoS in wireless multimedia sensor networks: A layered and cross-layered approach, *Wireless Personal Communications*, vol. 75, no. 1, pp. 729–757, 2014.
65. E. Gurses and O. Akan, Multimedia communication in wireless sensor networks, in *Proceedings of the Annals of Telecommunications*, vol. 60, pp. 872–900, 2005.
66. S. Misra, M. Reisslein, and G. Xue, A survey of multimedia streaming in wireless sensor networks, *IEEE Communications Surveys & Tutorials*, vol. 10, pp. 18–39, 2008.
67. M.M. Mohammad Masdari and M.T. Maryam Tanabi, Multipath routing protocols in wireless sensor networks: A survey and analysis, *International Journal of Future Generation Communication and Networking*, vol. 6, no. 6, pp. 181–192, 2013.
68. S. K. Singh, T. Das, and A. Jukan, A survey on internet multipath routing and provisioning, *IEEE Communications Surveys & Tutorials*, vol. 17, pp. 2157–2175, 2015.
69. P. Goyal, X. Guo, and H. M. Vin, A hierarchical CPU scheduler for multimedia operating systems, in *Proceedings of the 2nd USENIX Symposium on Operating System Design and Implementation*, Seattle, WA, pp. 491–505, October 1996.
70. S.C. Nelson et al., Understanding and developing a dynamic manycast solution for DTNs, 2011.
71. I. Lee et al., Wireless multimedia sensor networks, *Guide to Wireless Sensor Networks*, pp. 561–582, 2009.
72. F. Akhtar and M.H. Rehmani, Energy replenishment using renewable and traditional energy resources for sustainable wireless sensor networks: A review, *Renewable and Sustainable Energy Reviews*, vol. 45, pp. 769–784, 2015.
73. N. Ostadabbasi, Analysis of routing algorithms for Energy harvesting wireless sensor network, 2013.

74. F.M. Al-Turjman, H.S. Hassanein, and M. Ibnkahla, Towards prolonged lifetime for deployed WSNs in outdoor environment monitoring, *Ad Hoc Networks*, vol. 24, Part A, pp. 172–185, 2015.

75. M.H. Anisi, G. Abdul-Salaam, M.Y.I. Idris, A.W.A. Wahab, and I. Ahmedy, Energy harvesting and battery power based routing in wireless sensor networks, *Wireless Networks*, pp. 1–18, 2015.

76. E. Setton, Y. Taesang, Z. Xiaoqing, A. Goldsmith, and B. Girod, Cross-layer design of ad hoc networks for real-time video streaming, *IEEE Wireless Communications*, vol. 12, pp. 59–65, 2005.

77. A.J. Goldsmith and S.B. Wicker, Design challenges for energy-constrained ad hoc wireless networks, *IEEE Wireless Communications*, vol. 9, pp. 8–27, 2002.

78. M. Abazeed et al., A review of secure routing approaches for current and next-generation wireless multimedia sensor networks, *Int. J. Distrib. Sen. Netw.*, vol. 2015, pp. 3–3, 2015.

79. P. Sinha, QoS issues in ad-hoc networks, in *Ad Hoc Networks*, P. Mohapatra and S. Krishnamurthy, Eds., Springer: US, pp. 229–247, 2005.

80. I.F. Akyildiz, T. Melodia, and K.R. Chowdhury, Wireless multimedia sensor networks: Applications and testbeds, *Proceedings of the IEEE*, vol. 96, pp. 1588–1605, 2008.

81. M.A. Hamid et al., Design of a QoS-aware routing mechanism for wireless multimedia sensor networks, in *Proceedings of the Global Telecommunications Conference, IEEE GLOBECOM*, pp. 1–6, 2008.

82. D. Chen and P.K. Varshney, QoS support in wireless sensor networks: A survey, in *Proceeding. International Conference. Wireless Networks (ICWN 04)*, CSREA Press: Athens, GA, pp. 227–233, June 2004.

83. W. Dargie and C. Poellabauer, *Fundamentals of wireless sensor networks: Theory and practice*, John Wiley & Sons: Chichester, UK, 2010.

84. G.T. Singh and F.M. Al-Turjman, A data delivery framework for cognitive information-centric sensor networks in smart outdoor monitoring, *Computer Communications*, vol. 74, pp. 38–51, January 2016.

85. D. Chen, P.K. Varshney, QoS support in wireless sensor networks: A survey, in *Proceedings of International Conference on Wireless Networks*, pp. 227–233, Las Vegas, NV, USA, June 2004.

86. M.A. Yigitel et al., QoS-aware MAC protocols for wireless sensor networks: A survey, *Computer Networks*, vol. 55, pp. 1982–2004, 2011.

87. G.T. Singh and F.M. Al-Turjman, Cognitive routing for Information-Centric sensor networks in Smart Cities, in *International Conference in Wireless Communications and Mobile Computing (IWCMC)*, pp. 1124–1129, 2014.

88. F. Al-Turjman and M. Gunay, CAR Approach for the Internet of Things, *Canadian Journal of Electrical and Computer Engineering*, vol. 39, pp. 11–18, 2016.

89. R.V. Kulkarni et al., Computational intelligence in wireless sensor networks: A survey, *IEEE Communications Surveys & Tutorials*, vol. 13, pp. 68–96, 2011.

90. R. Pabst et al., Relay-based deployment concepts for wireless and mobile broadband radio, *IEEE Communications Magazine*, vol. 42, pp. 80–89, 2004.

91. J. Jeong et al., Empirical analysis of transmission power control algorithms for wireless sensor networks, in *Proceedings of the Fourth International Conference on Networked Sensing Systems (INSS)*, pp. 27–34, 2007.

92. Kanellopoulos, D. High-speed multimedia networks: Critical issues and trends, in *Handbook of Research on Telecommunications Planning and Management for Business*, In Lee, Ed., IGI Global: Hershey, PA, USA, pp. 775–787, 2009.

93. Z. Bojkovic and D. Milovanovic, Challenges in mobile multimedia: Technologies and QoS requirements, in *Proceedings of the 7th WSEAS International Conference on Mathematical Methods and Computational Techniques In Electrical Engineering, World Scientific and Engineering Academy and Society (WSEAS): Sofia, Bulgaria*, pp. 7–12, 2005.

94. S. Aswale and V. R. Ghorpade, Survey of QoS routing protocols in wireless multimedia sensor networks, *Journal of Computer Networks and Communications Article ID 824619*, vol. 2015, p. 29, 2015.

95. S. Md Zin, N. Badrul Anuar, M. L. Mat Kiah, and I. Ahmedy, Survey of secure multipath routing protocols for WSNs, *Journal of Network and Computer Applications*, vol. 55, pp. 123–153, 2015.

96. D. Goyal and M. R. Tripathy, Routing protocols in wireless sensor networks: A survey, in *Second International Conference on Advanced Computing & Communication Technologies (ACCT)*, pp. 474–480, 2012.

97. M. Gupta, N. Kumar, Node-disjoint On-demand multipath routing with route utilization in ad-hoc networks, *International Journal of Computer Applications*, vol. 70, no. 9, pp. 29–33, 2013.

98. P. Jian, E. Manning, and G. C. Shoja, Routing reliability analysis of partially disjoint paths, in *IEEE Pacific Rim Conference Computers and signal Processing (PACRIM)*, vol. 1, pp. 79–82, 2001.

99. M. Tarique et al., Survey of multipath routing protocols for mobile ad hoc networks, *Journal of Network and Computer Applications*, vol. 32, pp. 1125–1143, 2009.

100. T. Watteyne et al., From MANET To IETF ROLL standardization: A paradigm shift in WSN routing protocols, *IEEE Communications Surveys & Tutorials*, vol. 13, pp. 688–707, 2011.

101. S. Mueller et al., Multipath routing in mobile ad hoc networks: Issues and challenges, in *Performance Tools and Applications to Networked Systems*, vol. 2965, M. Calzarossa and E. Gelenbe, Eds., Springer: Berlin, Germany, pp. 209–234, 2004.

102. S. Adibi and R. Jain, Quality of Service Architectures for Wireless Networks, Performance Metrics and Management, 2010.

103. G. Spanogiannopoulos et al., A simulation-based performance analysis of various multipath routing techniques in zigBee sensor networks, in *Ad Hoc Networks*, vol. 28, J. Zheng et al., Eds., Springer: Berlin, Germany, pp. 300–315, 2010.

104. M. Abazeed et al., A review of secure routing approaches for current and next-generation wireless multimedia sensor networks, *International Journal of Distributed Sensor Networks*, vol. 2015, p. 3, 2015.

105. V. Bhandary, A. Malik, and S. Kumar, Routing in wireless multimedia sensor networks: A survey of existing protocols and open research issues, *Journal of Engineering*, vol. 2016, p. 27, 2016.

106. T. Cevik, A. Gunagwera, and N. Cevik, A survey of multimedia streaming in wireless sensor networks: Progress, issues and design challenges, *International Journal of computer Networks & Communications*, vol. 7, pp. 95–114, 2015.

107. J. Chen, M. Diaz, L. Llopis, B. Rubio, and J. M. Troya, A survey on quality of service support in wireless sensor and actor networks: Requirements and challenges in the context of critical infrastructure protection, *Journal of Network and Computer Applications*, vol. 34, pp. 1225–1239, 2011.

108. A. Jayashree et al., Review of multipath routing protocols in wireless multimedia sensor network: A survey, *International Journal of Scientific & Engineering Research Vohmm*, vol. 3, no. 7, pp. 1–9, 2012.

109. S. Gurung and D. Saikia, A survey of multipath routing schemes of wireless mesh networks, *International Journal of Computer Applications*, vol. 125, no. 14, pp. 12–20, 2015.

110. N.S. Nandiraju et al., Multipath routing in wireless mesh networks, in *IEEE Proceeding Conference International on Mobile Adhoc and Sensor Systems (MASS)*, pp. 741–746, 2006.

111. D. Ganesan et al., Highly-resilient, energy-efficient multipath routing in wireless sensor networks, *SIGMOBILE Mobile Computing and Communications Review*, vol. 5, pp. 11–25, 2001.

112. J. Guo et al., A Cross-layer and multipath based video transmission scheme for wireless multimedia sensor networks, *Journal of Networks*, vol. 7, no. 9, pp. 1334–1340, 2012.

113. K. Sha et al., WEAR: A balanced, fault-tolerant, energy-aware routing protocol in WSNs, *International Journal of Sensor Networks*, vol. 1, pp. 156–168, 2006.

114. L. Gan, J. Liu, and X. Jin, Agent-Based, Energy Efficient Routing in Sensor Networks, in *Proceedings of the Third International Joint Conference on Autonomous Agents and Multiagent Systems*, New York, vol. 1, pp. 472–479, 2004.

115. A. A. Ahmed and N. Fisal, A real-time routing protocol with load distribution in wireless sensor networks, *Computer Communications*, vol. 31, pp. 3190–3203, 2008.

116. K. Akkaya and M. Younis, Energy-aware delay-constrained routing in wireless sensor networks, *International Journal of Communication Systems*, vol. 17, pp. 663–687, 2004.

117. M. S. Al-Fares et al., A reliable multihop hierarchical routing protocol in wireless sensor network (WSN), in *ITNG '09. Sixth International Conference on Information Technology: New Generations*, pp. 1604–1605, 2009.

118. Y. Ming Lu and V. W. S. Wong, An energy-efficient multipath routing protocol for wireless sensor networks, *International Journal of Communication Systems*, vol. 20, pp. 747–766, 2007.

119. D. Goyal and M.R. Tripathy, Routing protocols in wireless sensor networks: A survey, in *Second IEEE International Conference on Advanced Computing & Communication Technologies (ACCT)*, pp. 474–480, 2012.

120. I. F. Akyildiz, W. Su, Y. Sankarasubramaniam, and E. Cayirci, Wireless sensor networks: A survey, *Computer Networks*, vol. 38, pp. 393–422, 2002.

121. X. Liu, Atypical hierarchical routing protocols for wireless sensor networks: A review, *IEEE Sensors Journal*, vol. 15, pp. 5372–5383, 2015.

122. X. Mande and G. Yuanyan, Multipath routing algorithm for wireless multimedia sensor networks within expected network lifetime, in *Proceedings of the International Conference on Communications and Mobile Computing (CMC)*, pp. 284–287, 2010.

123. Y. Sun, X.J. Shen, H.P. Chen, Energy balancing multipath routing protocol in wireless multimedia sensor networks, *Applied Mechanics and Materials*, vols. 155–156, pp. 245–249, 2012.

124. J. Guo, L. Sun, and R. Wang, A Cross-layer and multipath based video transmission scheme for wireless multimedia sensor networks, *Journal of Networks*, vol. 5, no. 6, pp. 1334–1340, 2012.

125. K. Mizanian et al., RETRACTED: Worst case dimensioning and modeling of reliable real-time multihop wireless sensor network, *Performance Evaluation*, vol. 66, pp. 685–700, 2009.

126. S. Dulman, T. Nieberg, J. Wu, and P. Havinga, Trade-off between traffic overhead and reliability in multipath routing for wireless sensor networks, in *Proceedings of the IEEE Wireless Communications and Networking Conference (WCNC)*, vol. 3, pp. 1918–1922, 2003.

127. N. Nasser and Y. Chen, SEEM: Secure and energy-efficient multipath routing protocol for wireless sensor networks, *Computer Communications*, vol. 30, pp. 2401–2412, 2007.

128. B. Yahya and J. Ben-Othman, REER: Robust and energy efficient multipath routing protocol for wireless sensor networks, in *Proceedings of the IEEE Global Telecommunications Conference, (GLOBECOM)*, pp. 1–7, 2009.

129. K. Akkaya and M. Younis, An energy-aware QoS routing protocol for wireless sensor networks, in *Proceedings. 23rd International Conference on Distributed Computing Systems Workshops*, pp. 710–715, 2003.

130. K. Lin et al., Adaptive reliable routing based on cluster hierarchy for wireless multimedia sensor networks, *EURASIP Journal of Wireless Communications and Networking*, vol. 2010, pp. 1–13, 2010.

131. G. H. EkbataniFard et al., A multi-objective genetic algorithm based approach for energy efficient QoS-routing in two-tiered wireless sensor networks, in *Proceedings of the 5th IEEE International Symposium on Wireless Pervasive Computing (ISWP)*, Mondena, Italy, pp. 80–85, 2010.

132. M. Liu, S. Xu, and S. Sun, An agent-assisted QoS-based routing algorithm for wireless sensor networks, *Journal of Network and Computer Applications*, vol. 35, pp. 29–36, 2012.

133. S.-L. Wu and Y.-C. Tseng, *Wireless Ad Hoc Networking: Personal-Area, Local-Area and the Sensory-Area Networks*, Auerbach Publications, Boca Raton, FL, USA, 2007.

134. N. Bulusu and S. Jha, *Wireless Sensor Network Systems: A Systems Perspective*, (Artech House Mems and Sensors Library), vol. 326, 2005.

135. J. Kulik, W. Heinzelman, and H. Balakrishnan, Negotiation-based protocols for disseminating information in wireless sensor networks, *Wireless Networks*, vol. 8, pp. 169–185, 2002.

136. D. Braginsky and D. Estrin, Rumor routing algorthim for sensor networks, *Proceedings of the 1st ACM International Workshop on Wireless Sensor Networks and Applications*, Atlanta, GA, USA, 2002.

137. J. Faruque and A. Helmy, Gradient-based routing in sensor networks, *SIGMOBILE Mobile Computing and Communications Review*, vol. 7, pp. 50–52, 2003.

138. S. Li et al., EDGE: A routing algorithm for maximizing throughput and minimizing delay in wireless sensor networks, in *IEEE Conference in Military Communications (MILCOM)*, pp. 1–7, 2007.

139. R. K. N. Shuang Li, C. Liu, A. Lim, Efficient multipath protocol for wireless sensor networks, *International Journal of Wireless & Mobile Networks (IJWMN)*, vol. 2, no. 1, pp. 110–130, 2010.

140. S. Chien-Chung et al., Sensor information networking architecture and applications, *IEEE Personal Communications*, vol. 8, pp. 52–59, 2001.

141. M. Abolhasan, T. Wysocki, and E. Dutkiewicz, A review of routing protocols for mobile ad hoc networks, *Ad Hoc Networks*, vol. 2, pp. 1–22, 2004.

142. R. Leung et al., MP-DSR: A QoS-aware multipath dynamic source routing protocol for wireless ad-hoc networks, in *Proceedings. LCN 2001. 26th Annual IEEE Conference on Local Computer Networks*, pp. 132–141, 2001.

143. J. W. Stefan Dulman, and P Havinga, *An Energy Efficient Multipath Routing Algorithm for Wireless Sensor Networks,*Enschede, the Netherlands: University of Twente, Centre for Telematics and Information Technology (CTIT), p. 6, 2003.

144. R.C. Shah and J.M. Rabaey, Energy aware routing for low energy ad hoc sensor networks, in *Conference in Wireless Communications and Networking WCNC 2002.* IEEE, vol. 1, pp. 350–355, 2002.

145. P. Hurni and T. Braun, Energy-efficient multipath routing in wireless sensor networks, in *Ad-hoc, Mobile and Wireless Networks.* vol. 5198, D. Coudert et al., Eds., Springer: Berlin / Heidelberg, Germany, pp. 72–85, 2008.

146. P. Hurni and T. Braun, Evaluation of wiseMAC on sensor nodes, in *Wireless and Mobile Networking.* vol. 284, Z. Mammeri, Ed., Springer: Boston, MA, pp. 187–198, 2008.

147. E.M. Royer and T. Chai-Keong, A review of current routing protocols for ad hoc mobile wireless networks, *IEEE Personal Communications,* vol. 6, pp. 46–55, 1999.

148. S. Sesay, Z. Yang, and J. He, A survey on mobile ad hoc wireless network, *Information Technology Journal,* vol. 3, pp. 168–175, 2004.

149. R. Benlamri, Networked Digital Technologies, *Proceedings Part II: 4th International Conference (NDT),* Springer: Dubai, UAE, vol. 294, pp. 1–12 , April 24–26, 2012.

150. C. Mbarushimana and A. Shahrabi, Comparativestudy of reactive and proactive routing protocols performance in mobile Ad Hoc networks, in *21st International Conference on Advanced Information Networking and Applications Workshops (AINAW),* pp. 679–684, 2007.

151. S. Lipsa, An empirical study of multipath routing protocols in wireless sensor networks, *International Journal of Computer Science & Information Technologies,* vol. 5, no. 4, pp. 5375–5379, 2014.

152. L. Schrage, *Optimization Modeling with LINGO,* LINDO Systems, Inc: Chicago, IL, 1998.

153. F. M. Al-Turjman, H. S. Hassanein, and M. Ibnkahla, Quantifying connectivity in wireless sensor networks with grid-based deployments, *Journal of Network and Computer Applications,* vol. 36, pp. 368–377, 2013.

154. M. H. Rehmani and A.-S. K. Pathan, *Emerging Communication Technologies Based on Wireless Sensor Networks: Current Research and Future Applications,* CRC Press, 2016.

155. S. H. R. Bukhari, M. H. Rehmani, and S. Siraj, A survey of channel bonding for wireless networks and guidelines of channel bonding for futuristic cognitive radio sensor networks, *IEEE Communications Surveys & Tutorials,* vol. 18, pp. 924–948, 2016.

156. O. B. Akan, O. B. Karli, and O. Ergul, Cognitive radio sensor networks, *IEEE network,* vol. 23, pp. 34–40, 2009.

157. A. O. Bicen, V. C. Gungor, and O. B. Akan, Delay-sensitive and multimedia communication in cognitive radio sensor networks, *Ad Hoc Networks,* vol. 10, pp. 816–830, 2012.

158. A. Fallahi and E. Hossain, QoS provisioning in wireless video sensor networks: A dynamic power management framework, *IEEE Wireless Communications,* vol. 14, pp. 40–49, 2007.

159. K. TariqJamilSaifullah and S. MuhammadFarrukh, Realizing Cognitive Radio Technology for Wireless Sensor Networks, in *Emerging Communication Technologies Based on Wireless Sensor Networks,* CRC Press, pp. 259–272, 2016.

160. F. M. Al-Turjman, Information-centric sensor networks for cognitive IoT: An overview, *Annals of Telecommunications,* pp. 1–16, 2016.

Chapter 3

Optimized Multi-Constrained Quality-of-Service Multipath Routing Approach for Multimedia Sensor Networks

3.1 Introduction

The next evolution of the sensing network paradigm is emerging toward wireless mobile sensor networks (WMSNs) due to numerous evolving multimedia real-time applications that depend on real-time communications [1]. Typical WMSNs consist of a large number of multi-functional, low-cost, low-power nodes made possible due to advances in micro-electro-mechanical systems (MEMSs) [2]. These sensor nodes are equipped with inexpensive miniaturized hardware, such as complementary metal-oxide semiconductor (CMOS) cameras, microphones, and other embedded components [3], which necessitates an efficient multipath routing approach in order to deliver their multimedia contents in a timely manner while satisfying quality of services (QoS) requirements [4].

Several approaches have been proposed in the literature for QoS support in WMSNs [3,4]. These approaches proposed various routing metrics, such as the shortest path, link stability, and minimum number of hops toward the sink [7]. Indeed, these approaches play an invaluable role in improving the realism of routing protocols; however, they do not provide a systematic way to generalize the analytical model beyond specific environments and radio conditions [7].

Accordingly, it is important for each sensor node in the network to consider the resource allocation process as an optimization problem with different potential goals [9]. This means that a sensor should consider resource utilization from a perspective of need, that is, the hop information. Furthermore, the resource allocation should be considered from a global perspective in which the utilization of resources is considered by all of the sensor nodes [5]. Consequently, a key issue in the design of a routing protocol is to determine the optimal amount of resources to be allocated, which necessitates the development of systematic mechanisms [1]. One of these mechanisms is a multi-constraint QoS routing, which is considered a key solution for QoS support in multimedia applications in which efficient routing strategies are needed for selecting energy-efficient paths for QoS requirements [10]. The multi-constraint QoS-routing determines the optimal level of resources that must be allocated to each selected path in terms of power consumption, delay, and throughput; thus, QoS requirements can be satisfied [11,12].

In this chapter, we investigate how to balance the use of resources and transmission radio to provide optimal QoS parameters as well as to avoid the overuse of the network resources. We prove the efficiency of considering multi-objective function analysis for adaptive switching in hop-by-hop QoS parameters. We assume that the analysis and design solutions for multimedia transmission in WMSNs require a high data rate with a low bit error probability (BEP) under certain channel conditions. The calculation of the BEP and the signal-noise ratio (SNR) in most multimedia applications of WMSNs are based on path loss propagation, which impacts QoS parameters [11]. There are many other factors that impact the results of QoS parameters for WMSNs; thus, a combination of optimization and analysis is required [6]. Toward this end, our main contributions are summarized as follows:

■ We propose a new routing metric for optimizing the link quality (LQ) between the sensor nodes thats translates into an efficient path selection mechanism. This mechanism uses mixed integer programming (MIP), which is based on the Lagrangian relaxation (LR) method, to define critical parameters to control the adaptive switching of the hop-by-hop QoS routing protocol. The embedded criteria for each objective function that is related to the decision constraints are used to decide which path from the source to the sink will be selected.

■ To highlight the novelties of our proposed approach for optimized, multi-constrained QoS-routing of WMSNs, the results of the proposed model are discussed in comparison to existing models. Extensive simulations demonstrate the robustness of our proposed routing protocol, which is compared to the performance of other popular routing protocols, such as reliable information forwarding using multiple paths (Re-InForM) [13], the multi-constraints multipath (MCMP) routing protocol [14], and the energy-constrained multipath (ECMP) protocol [15] using several performance metrics such as power consumption, average delay, and packet received ratio (PRR).

3.2 Related Works

Multipath routing protocols in the literature can be classified into single- versus multi-constrained approaches. In single-constrained approaches, there is mainly one parameter that controls the data delivery mechanism, such as reliability in Re-InForM [13] and energy in ECMP [15]. However, in multi-constrained routing protocols, several parameters can control the data delivery process. For example, in Reference 14, a distributed approach called MCMP is utilized in providing an end-to-end QoS-based data delivery. MCMP addresses the issues of multi-constrained QoS in WMSNs through a combination of the multipath traffic resources and the sensor node communication interface. It relies on obtaining both the additive delay and the multiplicative reliability of the selected route. Meanwhile, the Re-InForM protocol employs a probabilistic flooding to deliver an information awareness packet and service at desired priority levels of reliability. Unfortunately, this protocol is not designed specifically for real-time or multimedia traffic; therefore, it does not consider the delay deadlines of the packet when selecting the multiple paths. In Reference 15, the ECMP approach attempts to find a subset from a set of sensor nodes that have a lower expected energy transmission while meeting the QoS requirements by delivering traffic control packets to the sink. The ECMP approach searches for the subset of multipaths from the source to the sink that satisfies the requirements of the data source for the QoS and the total energy of transmission. Unlike our proposed mathematical framework, which is borrowed from MIP and is based on the LR method, we define critical parameters to control the adaptive switching of hop-by-hop QoS routing protocols.

To summarize, previous improvements in routing protocols were applied to address the computational complexity without implementing an analytical model for the time-varying network performance, which significantly impacts the multi-constrained QoS parameters due to temporal and spatial time complexity and/or space complexity. Unlike previous approaches, our proposed routing protocol formulates the problem of finding a path subject to multi-constraint routing in an analytical way. This is achieved by designing a new routing algorithm for traffic engineering that implements capacity provisioning based on the partitioning of end-to-end QoS parameters using MIP. The proposed routing protocol is, to the best of our knowledge, a first step toward QoS routing implementation in real-world tested platforms.

3.3 Partitioning Multi-Constrained Multipath Routing (PMMR) Protocol

Our proposed routing protocol uses MIP, which defines a critical parameter for solving nondeterministic polynomial time problems (NP-problems). The protocol is motivated by the need to find a scheme to increase the capacity of multi-service

Internet protocol networks [12]. The inspiration for the methodology of multi-service protocols for traffic engineering is to design a new routing metric that implements capacity provisioning based on the partitioning of end-to-end QoS parameters by periodically collecting the information between a source node and the next hop neighbor in the network. Our goal is to determine the optimal path that satisfies all of the QoS requirements for an efficient routing protocol over a multi-hop route.

3.3.1 Problem Formulation

The problem of energy efficiency for the application of embedding sensors in road-beds can be modeled with a directed network topology $G = (V, E)$, where $|V| = n$ is the number of sensor nodes and $|E| = l$ denotes the set of links in the network. Each node is characterized by a transmission range. Moreover, each sensor node enters a sleep state once there is no ongoing transmission; otherwise, it enters a wake-up state. The existing link between two sensor nodes is defined as $e = (\xi_i, \xi_j)$ from node ξ_i to node ξ_j, where $i = j = 1, ..., n$. We assume that each link $e \in E$ is characterized by two integer values: the energy consumption and the delay. Furthermore, the link e energy/delay functions are considered to be objective functions. We assume that there are $\xi_{i,j}$ possible routes between each pair $(i, j \in V)$ and a decision variable $X_v^{i,j}$ defined as a variable that has the value 1 if there is connectivity between two sensor nodes and has the value 0 otherwise. All proposed routing variables with their acronyms and descriptions are presented in Table 3.1.

3.3.2 Link Quality Modeling

Sensor nodes in wireless networks send information to each other by sharing common media. The issue is how to organize the sensor nodes in the network topology to obtain as much real-time information as possible [16]. To address this issue, a new link quality (LQ) model is needed based on empirical data (since the LQ becomes a common problem in wireless communications) while considering factors affecting energy consumption and QoS parameters [17]. Therefore, we use a combination of analytical and empirical methods to analyze common characteristics of the time-varying channel conditions [18,19]. The statistical parameters that define the transmission channel vary slowly with respect to the interval of successful data packet transmission, which complicates the performance analysis for the upper-layer protocols [20]. Consequently, the alternative solution is to approximate and model the fading channel as a discrete memoryless, finite-state, two-state Markov chain that is widely acknowledged as a reasonably accurate and mathematically tractable approach, which was first employed in Reference 21 (as shown in Figure 3.1) and is characterized by a specific packet error rate (PER). This two-state modeling is proposed as a probabilistic process by using a Bernoulli distribution for distributing consecutive packets and the signal-bit error correction mechanism into a finite number of level-crossing rates for SNR,

Table 3.1 Terminology Used for MIP Formulation

Variable	Description
n	Number of sensor nodes
$G = (V, E)$	Directed graph that represents network topology
ξ	Denoted to the sensor node
$X_v^{i,j}$	Defined as a decision variable
ξ	Sensor node
$\pm\mu$	Defined as a vector of LR of adding and subtracting the equality constraints
Γ	Link quality
P_{gb}	Transition probabilities from good state g to bad state b
π_0	The steady state of channel
λ	Number of transmission packets
χ_e	Packet error rate (PER)
L_p	Bits of data frame
γ	The received instantaneous SNR
P_ψ	Random variable
PRR	Packet received ratio
BER	Bit error rate
E_{elec}	The overhead energy due to the sensing, receiving, and processing
ε_{fs}	The loss coefficient related to p bit transmission propagated over the single-path model
ε_{mp}	The loss coefficient related to p bit transmission propagated over the multipath model
E_{tx}	The amount energy needed to transmit
E_{rx}	The amount of energy needed to receive
P_t	The transmitted power
P_n	The noise floor

(Continued)

Table 3.1 (*Continued*) Terminology Used for MIP Formulation

Variable	Description
ρ	The encoding ratio
σ	Random variable drawn from a Gaussian distribution with zero-mean κ
η	Number of hops
$alink_v^{ij}$	Indicator the partitioned link lies on the selected optimal path
d	The distance between sensor nodes
$\wp$	The end-to-end delay
$L_e^\wp$	The hop delay requirement along the path
$\Delta_\wp$	The bounded delay
Z	Dualizing the objective function with obtaining constraint

denoted as Γ to derive a mathematical model for the LQ, which is called a good channel and a bad channel, denoted as 0 and 1, respectively. Consider mapping a finite number of level-crossing rates as the tth state, in which an SNR that falls between $\Gamma_{(g_t)}$ and $\Gamma_{(g(t+1))}$ represents a good state g. The bad channel state b represents a LQ of the channel propagation situation in which achieving a proper reception of the data packets is extremely difficult. Each state is associated with transition probabilities P_{gb}, where $g, b \in 0, 1$ and $\sum P_{gb} = 1$ toggles from state g to state b, which is dependent on the propagation environment, transmission modulation scheme, and detection technique implemented at the receiver. We rely on an analytical approach to derive expressions for PRR as a function to analyze how the channel and radio determine the transitional region. In other words, the channel transition probability matrix considering different links experiences different levels of fading as

$$channel\ transition = \begin{bmatrix} p_{00} & p_{01} \\ p_{10} & p_{11} \end{bmatrix} = \begin{bmatrix} p & 1-p \\ 1-BER & PRR \end{bmatrix}. \tag{3.1}$$

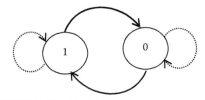

Figure 3.1 Two-state Markov chain transmission channel model.

For a Rayleigh fading channel with additive white Gaussian noise, the received instantaneous SNR γ is exponentially distributed as [18]

$$p(\gamma) = \frac{1}{\gamma_0} \exp\left(\frac{-\gamma}{\gamma_0}\right),$$ (3.2)

where γ_0 is the mean of the received SNR. The steady-state probability of a good channel state g_t is given by

$$\pi_0 = \int_{g_t}^{\Gamma_{g_t+1}} p(\gamma)d(\gamma) = \exp\left(-\frac{\Gamma_g}{\gamma_0}\right) - \exp\left(-\frac{\Gamma_g}{\gamma_0}\right).$$ (3.3)

Our mathematical model is used to derive the relationship between PRR and SNR and evaluates the trade-offs between them. For example, when the SNR value is greater than 20–50 dB, the PRR is almost 100%. A higher transmission power results in a higher signal-to-interference and SNR along the selected path but decreases the energy efficiency of the sensor node. Therefore, we determine the appropriate transmission power to reach the level-crossing rate of SNR at an appropriate value for a desired PRR by using a mathematical model that computes the optimal number of hops n for the source to the sink. To derive the transitional probabilities for consecutive packets λ assumed as a probabilistic process, assume P_ψ to be a random variable that is equal to 1 if the ψth packet is successfully received and that is equal to 0 otherwise; then, let the P_ψ be independent and identically distributed for $\psi = 1, \ldots, \lambda$. The expected value of P_ψ, $E(P_\psi)$ is equal to the probability that packet ψ is successfully received, that is, $P_\lambda(\psi)$. Hence, the average number of correctly received packets in the λ transmission is

$$PRR = \frac{1}{\lambda} \sum_{\psi=1}^{\lambda} P_\psi,$$ (3.4)

where $\psi = 1, 2, 3\ldots, \lambda$. Thus, for real-time applications in WMSNs in which λ is large, PRR can be closely approximated by the probability of a correctly received packet $E(P_\psi)$ [18]. Suppose that each sensor node knows the LQ of its neighboring nodes through χ_e, which is represented as the PER [22]. All of the PER calculations depend fundamentally on the bit error rate (BER) and SNR at the receiver. In order to formulate the optimization model, we must first analyze and solve the per-hop transmission packet between the source and the sink (node pair), either in direct or multi-hop communication under a specified per-hop PER; therefore

$$\chi_e = (1 - (1 - BER)^{L_p}),$$ (3.5)

where L_p represents the bits of data frame that need to be transmitted as $\tau = \dfrac{1}{(1-\chi_e)}$ (average retransmission rate), on the average for successful delivery of the packets. The receiver must estimate the noise and power received with knowledge about the modulation scheme, transmitted power, path loss, and noise floor in order to estimate the PER. Thus, the PRR can be closely approximated for multi-hop communication by

$$PRR \approx E(P_\psi) = (1-\chi_e)^\eta, \tag{3.6}$$

where

$$\chi_e = \frac{1}{\pi_0} \int_{\Gamma_{gt}}^{\Gamma_{gt}+1} \left(1 - \frac{1}{2} exp^{\left(-\frac{p(\gamma)}{2}\frac{1}{0.64}\right)}\right)^{\rho 8 L_p} d\gamma. \tag{3.7}$$

For each correctly received packet, L_p bits of data frame are received over a time period of $\dfrac{L_p}{bitrate}$, where

$$bitrate = numberofbitspersymbol * symbolrate. \tag{3.8}$$

Furthermore,

$$SNR = P_t - P_L(d) - P_n, \tag{3.9}$$

where P_t is the transmitted power in dB, and P_n is the noise floor of the receiver and antenna in dB that is given by

$$P_n = (F+1)kT_0B, \tag{3.10}$$

where F is the noise figure, k is the Boltzmann's constant, T_0 is the ambient temperature, and B is the equivalent bandwidth. Finally, P_L is represented as

$$P_L(d) = P_L(d_0) + 10\gamma \log\left(\frac{d}{d_0}\right) + \chi(0,\sigma), \tag{3.11}$$

where d is the distance between the source and sink, d_0 defines a reference distance between the two nodes, γ is the path loss exponent (which depends on the specific propagation environment), and $\chi(0, \sigma)$ is a random variable drawn from a Gaussian distribution with a zero-mean μ and a standard deviation of σ in dB [8]. A key issue in the calculation of the amount of power required by each node for its transmission is established by the power of the signal received at a node that is inversely proportional to d^α. This $P_r|d^\alpha$ relationship implies there is a trade-off between the power used by nodes and the number of hops in the path between communicating pairs of nodes if such path is selected. However, initial tests on the performance show the optimization of minimizing the energy consumption and the end-to-end delay exerted by increasing the probability of the link estimation. Putting a P_r, the SNR at distance d all together, and the PRR at a distance for the specific encoding and modulation is

$$PRR = \left(1 - \frac{1}{2}\exp^{-\frac{SNR}{2}\frac{1}{0.64}}\right)^{8L_p}.$$

(3.12)

To obtain a realistic radius of transitional region, we utilize the derivation reported in Reference 7 to determine the transitional region to bound the connected region to PRRs such that the transitional region values are bounded between 0.9 and 0.1:

$$SNR_{upper} = 10\log_{10}\left(-1.28\ln\left(2\left(1 - 0.9^{\left(\frac{1}{8L_p}\right)}\right)\right)\right)$$

(3.13)

and

$$SNR_{lower} = 10\log_{10}\left(-1.28\ln\left(2\left(1 - 0.1^{\left(\frac{1}{8L_p}\right)}\right)\right)\right).$$

(3.14)

Incorporoting Equation 3.11, then for a given the transmitted power P_t, the received power P_r at a distance d is bounded by [8]

$$Power_{Receupper} = P_t - \overline{PL(d)} + 2\sigma.$$

(3.15)

3.3.3 Neighboring Node-Disjointed Discovery Procedure

Each node invokes the discovery procedure by periodically broadcasting a HELLO message to its neighboring nodes in order to have enough information about which next hop node satisfies the forwarding condition. During this phase, each node can update and maintain its neighboring table. Some neighboring nodes receive the HELLO message and then send a reply message. Therefore, each node updates and maintains its neighboring table by recording or deleting the new neighbors. To construct and discover more multipaths, disjoint nodes modify their hop distance field in a HELLO message and look for better routing paths according to the updated information. As a result, each node becomes aware of its hop distance

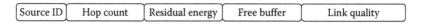

| Source ID | Hop count | Residual energy | Free buffer | Link quality |

Figure 3.2   The structure of a HELLO message.

from its neighboring nodes. Figure 3.2 illustrates the structure of a HELLO message, whereas source ID identifies the source. The hop count expresses the distance between nodes and its neighboring in hop for data transmission. Therefore, set a hop count variable to 1 and broadcast the HELLO packet further, updating the hop-count field to its own. A node that receives multiple HELLO messages from neighboring nodes updates its own hop-count variable to the least values received plus 1, and it keeps forwarding the message (with the updated hop count). Residual energy unveils the remaining energy in the battery of the node. Free buffer indicates the size of the buffer in the node. Finally, the LQ field is expressed in terms of SNR (that is, the LQ between the node and its neighbors).

3.3.4 Path-Disjointed Discovery Procedure

The sink node begins with a multipath discovery phase to create a group of neighbors that is able to forward the data packet toward the sink from the source sensor node. The constructed multipath is a node-disjoint path, which is usually preferred because it utilizes most of the network resources. This means that only the intermediate sensor nodes that are aligned on the partitioned selected path are awoken to forward the data packet toward the sink as depicted in Figures 3.3 and 3.4.

3.3.5 Path-Disjointed Selection Procedure

After completing the discovery phase and after all multiple paths have been constructed, the mechanism of path selection starts to select a set of multiple paths from the constructed paths to transfer the data packet from the source toward the sink. The selection phase of partitioning the multipath is based on a routing metric in order to optimize the selected paths in terms of minimizing the power

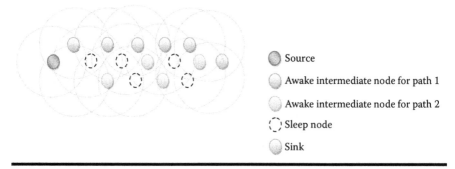

Figure 3.3   The configuration of network topology.

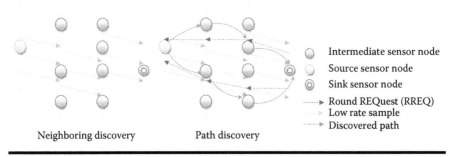

Neighboring discovery   Path discovery

Figure 3.4   Illustrating the construction of neighboring and path discovery.

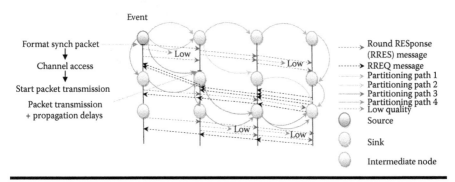

Figure 3.5   Description of PMMR protocol.

consumption and end-to-end average delivery delay. However, the selection is based on the definition of critical parameters to control the adaptive switching of hop-by-hop until the sink node, as illustrated in Figure 3.5. The path selection is based on partitioning paths, whereas the sensor nodes are distributed into partitioning different deployments. All sensor nodes inside the partitioned area can communicate with each other. Therefore, all sensor nodes in each partitioned area are equivalent for multipath routing in terms of quality of the link. The partitioning of the multipath is to optimize performance metrics such as energy balance and network lifetime. After constructing all multiple paths and after the source node receives data along these multiple paths, the source starts to select an optimal path from its next most-preferred neighbor.

3.4 Multi-Constraints QoS Parameters Modeling

QoS mapping is responsible for transforming the performances of QoS requirements for higher network layers into other lower QoS parameter performances.

Therefore, any performance degradations of one parameter can degrade the performance of other parameters [23]. The relationship between QoS parameters in different layers in a typical communication stack must be defined to specify the architecture in which the QoS requirements can be provided in sensor networks. Our proposed model uses the physical layer to determine the LQ to the medium access control (MAC), as well as a network layer for metric routing. This metric routing is used as a reference for scheduling in the MAC and for controlling the adaptive switching of hop-by-hop QoS routing protocols in the network layer. Additionally, the duty cycle of the sensor node (wake up/idle/sleep/transmit/ receive) at the MAC layer controls the optimization of the multipath routing protocol. The optimization problem flow is modeled in the form of a series of energy and delay constraints presented in the MIP model for partitioning the multipath routing protocol.

3.4.1 Power-Consumption Modeling

We adopt the energy model in Reference 24 to derive a new mathematical model to minimize the power consumption, and then we determine the optimal number of hops. According to this model, the amount of energy to transmit L_p bits of information over η-hops along the selected path is [25]

$$E_{s_{ij}} = L_p \left\{ \sum_{i=j=1}^{\eta} 2[E_{elec_{ij}} + \varepsilon_{mp}(d_{ij})^\alpha] \right\}. \tag{3.16}$$

The objective function is to minimize the energy consumption for the sensor nodes, which are defined in Equation 3.16. The two variables that must be defined are the number of hops and the intermediate distance between two sensor nodes along the selected path. Thus, Equation 3.16 can be rewritten as

$$E_{\xi_{ij}} = L_p \left\{ \sum_{i=j=1}^{\eta} 2[E_{elec_{ij}} + \varepsilon_{mp}(d_{ij})^\alpha] \right\} X_v a_{link_{v_{ij}}}^{\xi_{ij}}. \tag{3.17}$$

The appropriate constraints are obtained from both, the number of hops from the source to the final sink and the intermediate distance between the two nodes. Because our application involves realistic WSN environments, it is necessary to find the optimal number of hops and the corresponding intermediate distance. The equation for a fixed intermediate distance is

$$\sum_{i=j=1}^{\eta} d_{ij} = d. \tag{3.18}$$

It must minimize the value of the energy of the selected path when $d_{11} = d_{22} = \cdots = d_{\eta} = \dfrac{total\,distance}{number\,of\,hops}$. Therefore, the optimal theoretical hop number is obtained as an integer number for the multipath model when the path loss exponent $\alpha = 4$ from

$$\eta^{optimal} = \sqrt[\alpha]{d\left(\frac{3\varepsilon_{mp}}{2E_{elec_{ij}}}\right)}. \tag{3.19}$$

Finally, the objective function for minimizing the energy for the partitioning path is

$$Z = \min E_{\xi_{ij}} = L_p \left[\sum_{i=j=1}^{\eta} 2[E_{elec_{ij}} + \varepsilon_{mp}(d_{ij})^{\alpha}] \right] x_v a_{link_{v_{ij}}}^{\xi_{ij}} \tag{3.20}$$

subject to

$$\eta^{optimal} = \sqrt[\alpha]{d_{ij}\left(\frac{3\varepsilon_{mp}}{2E_{elec_{ij}}}\right)} \sum_{v=1} x_v \leq d_{ij}, \tag{3.21}$$

$$x_v = \{0,1\}, \forall link \in E. \tag{3.22}$$

Equation 3.21 guarantees that the optimal number of hops between the selected paths can be obtained, whereas the second constraint defines a decision variable for the selection and partitioning. The optimization problem is solved by using the dualizing Equation 3.21 on the objective function using LR. A critical parameter is defined to control the adaptive switching of the hop-by-hop QoS-routing protocol. Thus, the embedded criteria based on the decision constraint used for each objective function decide the path from the source to the sink. Consequently, a_{link_v} indicates that the partitioned link lies on the selected optimal path; its value is 1 if the link lies on the selected path and is 0 otherwise. Therefore,

$$Z_d = \max_{\mu}\left\{ \min E_{\xi_{ij}}(\mu) = L_p\left[\sum_{i=j=1}^{\eta} 2[E_{elec_{ij}} + \varepsilon_{mp}(d_{ij})^{\alpha}] \right] \right.$$
$$\left. -\mu\left(\sqrt[\alpha]{d\left(\frac{3\varepsilon_{mp}}{2}E_{elec_{ij}}\right)}\right)\sum_{v=1} X_v - d_{ij} \right\}, \tag{3.23}$$

subject to

$$x_v = \{0,1\}, \forall link \in E, \tag{3.24}$$

where μ is defined as a vector of LR $\forall v$, where $v = 1, 2, 3, \ldots, m$

3.4.2 Delay-Constraint Modeling

The proposed routing algorithm involves multi-hops that increase the delay. Therefore, the proposed modeling assumes a LQ distributed metric on each route between two sensor nodes in terms of the optimal hop number, which could have different delay guarantees, denoted as $\wp(\xi_i,\xi_j)$. First, determine the optimal number of hops that minimizes the delay of the successful transmission of a packet, then jointly optimize the hops and the estimation of LQ to derive a scaling for minimizing the delay. The problem of determining the QoS route that satisfies the delay constraint is a difficult, NP-problem; it requires more information regarding nodes on their routes to the sink, which is impossible to obtain in WSNs [25]. Hence, to solve the optimization problem, all the source nodes and the intermediate nodes periodically calculate the LQ from the one-hop neighborhood of each node because the one-hop LQ is easier to acquire. Suppose that one QoS requirement is to be satisfied at each hop, then the end-to-end QoS requirement is also met. More precisely, a node can satisfy the hop requirement by selecting the next hop based on the steady state of the estimated LQ. This allows the bounded delay to be evenly divided at each hop, which is described as follows. The end-to-end delay between any two sensor nodes ξ_{Source} and ξ_{Sink} over the set of paths P is given by

$$\wp_{SourcesSink}(L_p) = \min\left\{\sum_{\xi_i}\wp(\xi_i,\xi_j)\right\}, \tag{3.25}$$

where $\wp_{SourcesSink}$ is the minimum achievable delay when the generated data are routed along the set of paths between ξ_{Source} and ξ_{Sink}. The delay $\wp(\xi_i,\xi_j)$ between two nodes is the time required to successfully transmit a packet after the first node receives it. This time might include queuing, contention, transmission, retransmission, and propagation. This approach is not similar to nonlinear programming, in which solving the problem depends on the end-to-end information that is acquired from the source node only. The approximate problem is solved by using MIP to partition the source and intermediate nodes and collect information at each hop to meet the overall QoS requirements and satisfy the constraints. Therefore, the problem is formulated as an integer programming problem, and the solution to this problem is provided by the LR method to uniformly partition the QoS requirements of all downstream optimal hops, which is formulated as follows. The primary integer programming problem

$$\wp_{SourcesSink} = \min\sum_{\xi \in P}\sum_{v=1}X_v\wp(\xi_i,\xi_j)a_{linkv}^{\xi_i\xi_j} \tag{3.26}$$

is subject to

$$\sum_{v=1}\wp(\xi_i,\xi_j) \leq X_v\Delta\wp, \tag{3.27}$$

where $\Delta\wp$ is the bounded delay, which depends on two factors: the number of hops taken and the delay of a node, which are of additive form and denoted as η_{ij} and $\wp^\epsilon$, respectively. Therefore,

$$\Delta\wp = \wp_0^{Source} + \wp_{\eta_1}^{\xi+1} + \wp_{\eta_2}^{\xi+2} + \cdots + \wp_{\eta_{hopssink}}^{Sink}. \tag{3.28}$$

Thus, $\Delta\wp$ can be rewritten as

$$\Delta\wp = \Sigma_{i=j=1}^{n}\wp^\epsilon + (n-1)\eta_{ij}. \tag{3.29}$$

This is the additive form of $\Delta\wp$. The partitioning QoS requirements are met at all downstream optimal hops. $L_e^\wp$ is the hop delay requirement along the path from the source to the sink, which is composed of η_i and depends on the partition requirements at sensor node ξ_i. The hop delay requirement is equal to

$$L_e^\wp = \frac{\Delta\wp - \wp^\epsilon}{\eta_i}. \tag{3.30}$$

Then

$$\sum_{v=1}\wp(\xi_i,\xi_j) \leq X_v L_e^\wp, \tag{3.31}$$

which expresses the delay of the selected route of the binary decision such that $X_v = 1$ if the delay is lower than the local delay requirement; otherwise, $X_v = 0$:

$$\sum_{Source}^{Sink} X_v = 1, \tag{3.32}$$

$$X_v = \{0,1\}. \tag{3.33}$$

There are two LR methods. The first is obtained from dualizing constraint 3.32,

$$Z_{\wp SourcesSink}(\mu) = \max_\mu\left\{\min \sum_{\xi\in P}\sum_{v=1} X_v\wp(\xi_i,\xi_j) + \mu\left(\Sigma_{Source}^{Sink} X_v - 1\right)\right\}, \tag{3.34}$$

subject to

$$\sum_{v=1} X_v\wp(\xi_i,\xi_j) \leq L_e^\wp, \tag{3.35}$$

$$X_v = \{0,1\}. \tag{3.36}$$

The second is obtained from dualizing constraint 3.33,

$$Z_{\wp SourcesSink}(\mu) = \max_{\mu} \left\{ \min \sum_{\xi \in P} \sum_{v=1} X_v \wp(\xi_i, \xi_j) + \mu(\Sigma_{v=1} X_v \wp(\xi_i, \xi_j) - L_e^\wp) \right\}, \quad (3.37)$$

subject to

$$\sum_{Source}^{Sink} X_v = 1, \quad (3.38)$$

$$X_v = \{0,1\}, \forall link \in E. \quad (3.39)$$

What follows is a little thought about using Equation 3.21, or Equations 3.22 and 3.32, or Equation 3.33 embedded into Equations 3.20 and 3.26 for finding the optimal solution or the near-optimal solution. The partitioning problem shows that different μ multiplier values can be derived for the same problem. However, comparing Equation 3.21, or Equations 3.22 and 3.32, or Equation 3.33, they are harder to solve but might provide better bounds since μ multiplier values are used to provide a good feasible area that includes the optimal solution of the optimization problem. Hence, the objective functions in Equations 3.20 and 3.26 have a number of important structural properties that make them feasible to solve. Consequently, the decision variables are also constrained to lie in a given set that is defined as the feasible solution area to a partitioning routing selection problem Ψ. Therefore, we assume that Ψ is finite as

$$\Psi = \left\{ X : \wp \mid E_{tx,ij} \le L_e^\wp \mid \eta^{optimal}, 0 \le X \le \mu \right\}. \quad (3.40)$$

The optimal values of μ help to determine the level of cutting-off of the feasible area that yields the optimal solution for the optimization problem.

3.5 Performance Evaluation

We evaluate the efficiency of the PMMR by comparing its performance against Re-InForM [13], MCMP [14], and ECMP [15], on the basis of energy consumption, average end-to-end delay, and PRR. The simulation investigates the effectiveness of the end-to-end QoS parameters of the PMMR protocol. We assume a total number of 50 sensor nodes distributed uniformly in an urban highway intended to detect abnormal events. Each sensor node is an acousto-magnetic radio frequency (AM RF) transceiver with a transmission range of 12.00 m. Among these sensor nodes, the probabilistic LQ of a discrete memoryless, finite-state, two-state Markov chain of channel reliability is examined. Whether or not the link quality achieves

Table 3.2 Definition of All Parameters in PMMR

Parameter	Value
E_{elec}	50 nJ/bit
ε_{fs}	10 pJ/bitm²
ε_{mp}	0.0013 pJ/bitm²
Topology structure	Urban highway, sensor node distributed uniformly
Number of sensor nodes	50 sensor nodes
Message payload	64 bytes
Data length p	2000 bits
Transmission range	12.00 m
Total distance	150 m

a good channel performance in these experiments is also investigated. Table 3.2 depicts parameters used in the PMMR simulation.

3.5.1 Experiment 1: Effectiveness of Cut-off Determination

The key to finding the optimal number of hops for a multipath from the source to the sink can be exposed to the convex envelope or the convex hull of the objective function by adding an objective function cut. Therefore, the evolution of energy consumption and end-to-end delay can be generated for the multipath that adopts the objective functions for the power consumption and delay level cut at an optimal number of hops corresponding to a feasible solution. The proponents of multi-hop communication routing argue that more short hops are preferable to fewer long hops because of the estimation of the LQ with a finite number of observations of distributed channel access. The PMMR protocol depends on a proposed LQ as a new routing metric for the forwarding mechanism in which every node makes a decision to route the packet to the closest neighbors or to the nodes that are closest to the sink. This routing metric is assumed to require a high data rate with a low BER, which is considered a factor for improving the amount of information transferred and for understanding such information in light of the design of a waveform under certain channel conditions. The simulation is performed with two different level cuts at 5 and 8 hops to indicate that it has met the QoS constraints, namely, the lower and upper constraints that correspond to a feasible solution for the optimization problem. The objective function for a path can be estimated by successively adding or subtracting the equality constraints to eliminate variables (i.e., surplus variables) and adding the ≤ inequality constraints in suitable nonnegative multiples. In other words, the objective function is expressed as a linear combination of the constraints to eliminate the variables. However, the estimation of energy consumption and end-to-end delay for a path that adopts the objective functions for the power consumption and delay can be verified through the level cuts at a specified number of hops. This estimation corresponds to a feasible solution to the constraints that, when added together in this multiple, will produce these constraint values. This mode is known as the dual mode by producing a Lagrangian problem that is easy to solve and whose optimal value is a lower bound on the optimal value of the objective function. In other words, the values of the energy consumption and end-to-end delay functions are 0.033 mW and 1.6 s, respectively, defining the upper bound of the possible values. Moreover, through having a weak duality, the values of the power consumption and end-to-end delay objective functions are 0.01 mW and 0.1 s, respectively, defining the lower bound.

The objective function value of the multiplier problem is always lower than that of the original problem. Therefore, adding an objective cut of 0.01 mW $\leq f(x) \leq$ 0.033 mW for the power consumption and $0.1s \leq f(x) \leq 1.6s$ for the end-to-end delay to the original problem encompass the optimal solution. The main original function seeking the optimal multiples on all constraints can be formulated to obtain the optimal solution. The values of the power consumption and end-to-end delay functions vary because of the different volumes of traffic environments. The traffic data flow is homogeneous since the behavior of vehicles passing alone an urban highway follows Poisson arrival. However, the proposed model investigates different degrees of periodicity in the traffic data flow in various environments, as depicted in Figure 3.6. Therefore, the upper and lower bound values for power consumption and end-to-end delay vary depending on the real-time traffic data flow in various environments, such as weekday, weekend, or morning and evening. The evolution of power consumption and end-to-end delay for the multipath adopts the level cut-off at the optimal number of hops corresponding to the generated feasible solution. The proponents of multi-hop communication routing argue that more short hops are preferable to fewer long hops because of the estimation of the LQ with a finite number of observations of distributed channel access [8]. Figure 3.7 shows the analytical PRR versus distance obtained from Equations 3.13 and 3.14, whereas the width of each region depends on the operational environment. We used the observation in Reference 8 in which the beginning and ending of the transitional region match the analytical values obtained as we analyzed the parameters chosen for our experiment (frame size $L(p) = 50$ bytes, $P_t = 0$ dBm, n is set to 4 and σ is bounded with 1, 2, and 4) we obtain $SNR_{upper} = 10.3565$ dB and $SNR_{lower} = 8.3941$ dB. Figure 3.8 shows that our proposed model estimates the influence the environment in different regions, which can be used to evaluate the performance of the routing protocol. Figure 3.8 illustrates the physical impact of these parameters in the width of the transitional region. The SNR bounds of the radio model are fixed and independent of the environment; hence, the P_r has a higher probability of entering the transitional region at a closer mean distance μ between sensor nodes when σ increases, leaving

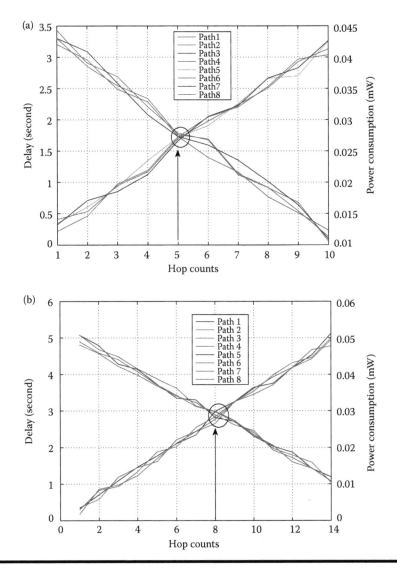

Figure 3.6 Adopting an objective function with the level cut method. (a) The level cut at 5 hops. (b) The level cut at 8 hops.

it at a farther distance, which results in a larger transitional region. We use the values of SNR as an immediate marker of LQ to present the results. Suppose that a node wakes up to monitor the urban highway at p random times, which are uniformly distributed throughout the duty cycle. Hence, the awakened node records the average SNR at that time, and then the recorded value is sufficient to obtain a good estimate of the link even for degraded links. Further, it can be observed

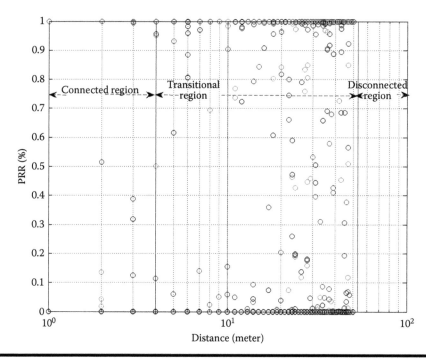

Figure 3.7 Analytical PRR for determining various regions of LQ.

that the SNR falls sharply whenever the distance decreases below 25 meters, which describes the interacting channel parameters to determine the extent of the connected, transitional, and disconnected regions. Therefore, the average SNR values along the partitioning routes must be determined to optimize the energy efficiency of the network topology by minimizing the energy consumption and delay. This goal is crucial because the SNR along the route is larger for multi-hop communication and accounts for the important practical issues of resource allocation, energy, and delay constraints. In the presence of two links, both links are affected by channel fading. The measured SNR values for the two links are 17 and 25 dB. The measured PRR is approximately 100% when the SNR is larger than 25 dB. The energy consumed by a link is strongly related to the status of the wireless channel among communicating nodes. This relation can be observed when the channel fading is on and affects both links similarly. The SNR of the former link decreases to 5 dB, and that of the latter remains at 17 dB. This result indicates that the latter has a more efficient LQ than the former.

Each node has a relatively high duty cycle to determine the recent state of the wireless channel to other nodes. The two links are compared in terms of their estimated SNR values. SNR is an important factor in refining the LQ judgment. This goal can be accomplished by increasing the SNR (the derivative of Equation 3.2 with respect to $p(\gamma)$) to weigh the calculated efficiency. The signal may be diffracted,

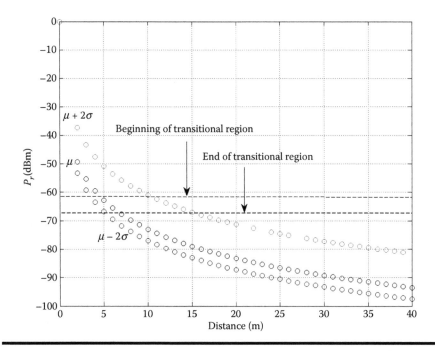

Figure 3.8   Interaction of channel parameters on determination of the transitional region.

scattered, and reflected. Therefore, these effects have two necessary consequences for the SNR. First, the SNR decays exponentially with respect to the distance. Second, for a given distance, the SNR is randomly and log-normally distributed about the mean distance-dependent value. The aim is to obtain the width of different regions, as shown in Figure 3.7. The PRR values significantly vary as the SNR values fluctuate from the beginning to the end of transitional region. However, the SNR value is determined on the basis of the distance between the partition nodes. The scatter plots of the measured PRR describe the physical interaction of the channel with the radio model conditioned on a finite number of level-crossing rates for the SNR. The possible PRR is then extracted from this SNR and is given by p_ψ.

3.5.2 Experiment 2: Comparison against Three Routing Algorithms

The amount of energy consumed in the transmission of data packets from a specific node to the sink or even to an intermediate node on the link is important for obtaining a good steady-state wireless channel between two partitioning nodes. Therefore, a routing algorithm based on statistical parameters that define the transmission channel is more efficient than a routing algorithm based on a straightforward hop count. To understand the behavior, consider that the probability of the wireless link estimation can control the reduction of the PER through the improvement of the SNR by decreasing the distance between two adjacent nodes. This construction deploys or wakes up a new node in the network and then increases the network density deployment. The increase in the LQ implicitly assumes that a new aligned node on the partitioning route has a relatively high duty cycle that can determine the recent state of a wireless channel to each node to transmit or receive packets. Other unaligned nodes implicitly save energy consumption by turning their radio off. Each node obtains LQ information about its neighbors. This information provides a basis for setting a solution for minimizing energy consumption. Figure 3.9 shows the influence of increasing the LQ on energy consumption when resolving the optimization problem by using the constraints of the transmission range, the optimal number of hops, and the integer decision variable. A single-hop transmission always consumes 10 times more energy than multi-hop communication does as the transmission range increases or decreases because the distance between the source and sink is fixed at $D = 150$ m. In multi-hop communication, the amount of energy consumed by the optimal aligned partitioned route globally decreases as the probability of the link efficiency estimation increases. Figure 3.10 shows the total power consumption in the network with four models that perform the multipath routing algorithm equally. The proposed model performs better than the ECMP and MCMP. This is attributed to the fact that both the ECMP and MCMP are deterministic zero-one problems with linear objective functions and constraints. The number of constraints can be expressesd as $2|N[\iota]| + 2$, where $|N[\iota]|$ defines the number of sensor nodes in the network. Therefore, the size of the search space increases with increasing node density. The optimal multipath becomes too large for acceptable computation time as the number of nodes increases. The Balas algorithm resolves the search method for multipath determination from the source to the sink that supports the QoS requirements. This algorithm is based on the random selection of the node at each next hop until the sink is selected. The ECMP model can achieve approximately 43.41% more power savings than the MCMP model because the latter uses 15.45% more multipath routing than the former. This means that the ECMO algorithm supports the QoS requirements in sensor networks based on well-defined QoS constraints to avoid any energy exhaustion that occurs when delivering the packets along the selected multipath. Unlike both models, the PMMR protocol routes the information over the optimal hop number from the source to the sink. The strength of the proposed model lies in the fact that it accomplishes a trade-off between the optimal hop number and the minimum energy consumption by selecting a path with the optimal hop number. The proposed model depends on a new search method for the selection of multipath routing based on the Lagrange multipliers μ for two nodes with an optimal hop number. Then, it transforms the constrained optimal path problem to the objective function of energy consumption (Equation 3.23) by using the level cut-off method. Therefore, this transformation for power consumption includes the

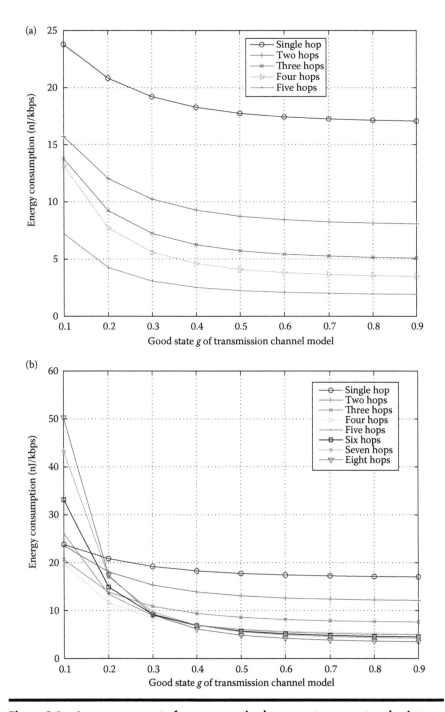

Figure 3.9   Average amount of energy required to report an event under integer optimization programming. (a) Level cut at five hops. (b) Level cut at eight hops.

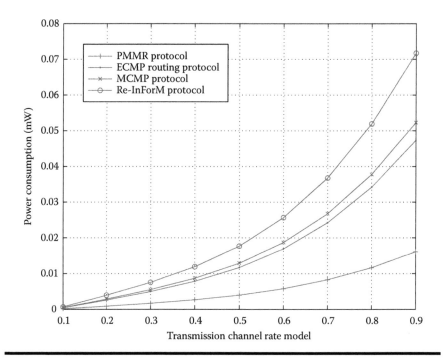

Figure 3.10 The power consumption comparison.

optimal solution (Figure 3.6). This solution enables the proposed model to achieve approximately 74.42% more power savings than the MCMP and ECMP models. The power consumption for Re-InForM for each successful packet received is given by [14]

$$N_s\Sigma_{i=0}^{k-1}(1-p_e)^i = \frac{(1-(1-p_e)^k)\log(1-r)}{e\log(1-(1-p_e)^k)}, \tag{3.41}$$

where k is the number of hops, and N_s is the number of transmission multipaths for each successfully received packet, which can be expressed as [14]

$$N_s = \frac{\log(1-r)}{\log(1-(1-p_e)^k)}, \tag{3.42}$$

where r defines the reliability of the transmission channel rate. Therefore, the Re-InForM protocol consumes more power than other models as $1 - p_e$ increases. The Re-InForM protocol demands more multipaths to transmit the packet with desirable reliability. Figure 3.11 shows that the benefits of multi-hop communication are eroded by the delay under delay constraint. It also shows the expected average data delivery delay from the source to the sink versus the probability of the link

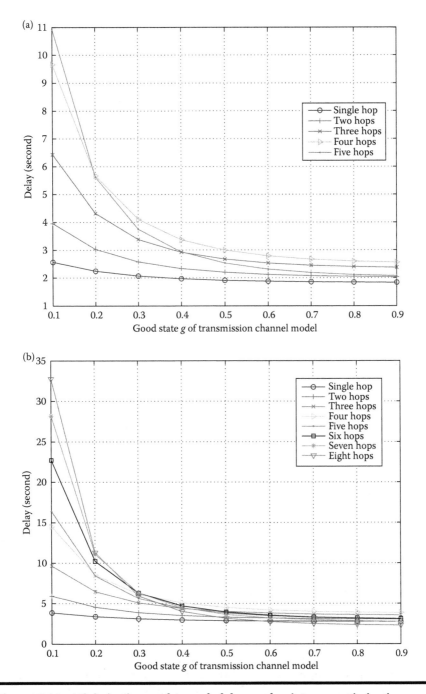

Figure 3.11 Minimization end-to-end delay under integer optimization programming. (a) The level cut at 5 hops. (b) The level cut at 8 hops.

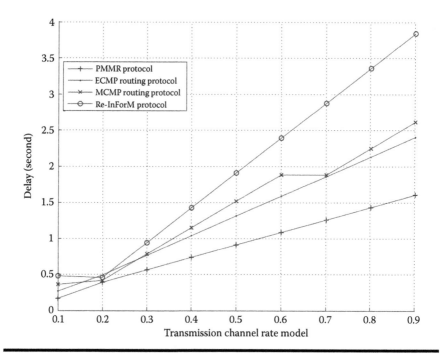

Figure 3.12   Average delivery end-to-end delay comparison.

quality estimate. The graph of the delay is a concave function, and a good steady state of the transmission channel model varies. As the delay shows almost the same pattern, a good state satisfies the constraints, and the minimum delay is a global optimization value.

Figure 3.11 demonstrates that the delay decreases when the packet travels along links with an estimated efficiency that connects the sensor nodes in the network domain. The probability that a link increases the sensor node deployment is high. As a result, the delay is minimized by applying the integer optimization problem constraint in which the end-to-end delay is constrained to find the optimal aligned partitioned route.

Moreover, the ECMP and MCMP models only exhibit slight differences in the average end-to-end delay. The multipaths selected by the two models have different numbers of hops (Figure 3.12). The ECMP model uses a smaller group of longer multipaths than the MCMP model. Therefore, the MCMP model consumes more power than the ECMP model, which lowers the average end-to-end delay. The similarity of the performances of the ECMP and the MCMP justifies that the trade-off between the power consumption and the average end-to-end delay is affected by the number of selected multipaths. This finding reveals that the ECMP model uses a smaller or longer multipath, resulting in lower power consumption with a higher average end-to-end delay than the MCMO model.

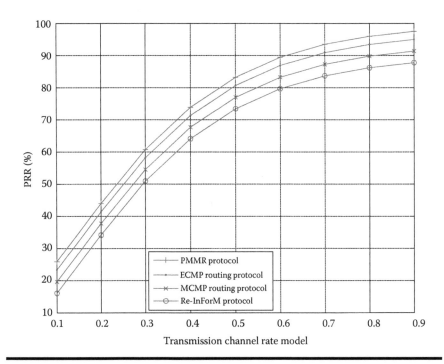

Figure 3.13   PRR comparison.

By contrast, our proposed model outperforms the other models by using its preferred selected multipath with an optimal hop number. This result reveals that the proposed model uses the multipath with the optimal hop number. Therefore, the proposed model is more likely to lead to a lower average end-to-end delay than the other two models examined in this chapter. The proposed model extends to the investigated network performance by studying the influence of the optimal hop number on the relationship between the SNR and the PRR. Figure 3.14 illustrates the relationship between the PRR and the SNR for various numbers of hops for the integer optimization problem constraint in which energy consumption is constrained. For the single hop, multi-hop, and optimal number of hops, the figure demonstrates that the optimal hop number along a selected partition path achieves a good PRR. The range of PRR in the experiments is 51%–90%. This result is the optimal rate for the optimal energy consumption with sufficiently strong signal reception strength, which has an SNR value ranging from 1–25 dB. The single hop of the selected partition path that consumes a high amount of energy has a PRR of 98.81%. The multi-hop of a selected partition path, PRR ranges from 81%–99%.

All models perform equally in terms of the PRR for the reliability of transmission channels (Figure 3.13). To have a close view of the PRR, the curves of all models display the PRR on a log scale. Among all the models, the PMMR protocol achieves the best PRR of approximately 96%. The ECMP, MCMP, and

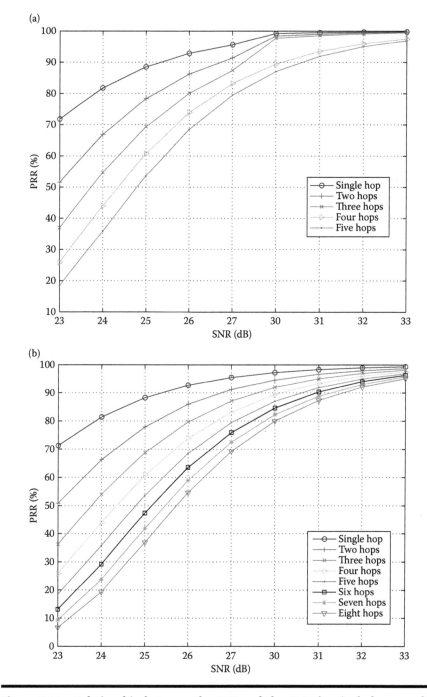

Figure 3.14 Relationship between the PRR and the SNR for single hops, multi-hops, and optimal hops. (a) The level cut at 5 hops. (b) The level cut at 8 hops.

Re-InForM models achieve PRR values of 94%, 92%, and 88%, respectively. The reception ratio of the packets usually increases logarithmically as the reliability of the transmission channel rate increases because of the confirmation of more packets that are successfully delivered to the sink, with a small expiring ratio for the lost packets. It is important to address the average cost of a routing algorithm in order to understand its computational complexity. The cost of a routing algorithm depends on searching the optimal selected path from the source to the sink. Usually, this depends on the number of the effective feasible solutions required to evaluate the objective functions in Equations 3.20 and 3.37, in terms of power consumption and delay, respectively, which counts to the desired search space of the optimization problem. And thus, the complexity is determined by the optimal value of μ required to obtain the optimal selected partitioning path as referred to in Equations 3.23, 3.34, and 3.37, with $Z_d = Z$. The average value of $Z_d * 100$ is divided by the average value of Z (Equations 3.20 and 3.37), where Z_{d^*} denotes the number of actual lower and upper bounds obtained from the determination of the cut-off.

3.6 Conclusions

One important conclusion from our proposed routing algorithm is that existing features of the multipath routing approach can be utilized to design a QoS approach for optimizing the QoS parameters in terms of power consumption, delay, and throughput for real-time applications. The partitioning multipath routing approach provides sufficient information about the links between sensor nodes to determine the optimal path and to select the intermediate sensor node for routing the packets from source to sink. The metric prioritizes the sensor nodes according to the factors of LQ based on a generic routing protocol. The results demonstrate that the proposed metric improves the PRR from the source to the sink, increasing the lifetime and minimizing the end-to-end delay. A future trend of our work will be focused on designing a new type of middleware software layer that stands between the network operating system and the real-time applications to provide solutions to frequently encountered heterogeneity, interoperability, and security issues.

References

1. A.A. Anasane and R.A. Satao, A survey on various multipath routing protocols in wireless sensor networks, *Procedia Computer Science*, vol. 79, pp. 610–615, 2016.
2. V. Bhandary, A. Malik, and S. Kumar, Routing in wireless multimedia sensor networks: A survey of existing protocols and open research issues, *Journal of Engineering*, vol. 16, no. 9, pp. 1438–1465, 2016.
3. I.F. Akyildiz, T. Melodia, and K.R. Chowdhury, A survey on wireless multimedia sensor networks, *Computer Networks*, vol. 51, pp. 921–960, 2007.

4. A.C. Fischer, et al., Integrating MEMS and ICs, *Microsystems & Nanoengineering*, vol. 1, pp. 15005–15021, 2015.
5. G. Singh and F. Al-Turjman, A data delivery framework for cognitive information-centric sensor networks in smart outdoor monitoring, *Elsevier Computer Communications Journal*, vol. 74, no. 1, pp. 38–51, 2016.
6. F. Al-Turjman, H. Hassanein, W. Alsalih, and M. Ibnkahla, Optimized relay placement for wireless sensor networks federation in environmental applications, *Wiley: Wireless Communication & Mobile Computing Journal*, vol. 11, no. 12, pp. 1677–1688, Dec. 2011.
7. W. Yang and Y. Zhang, A fast algorithm for the optimal constrained path routing in wireless mesh networks, *Journal of Communications*, vol. 11, no. 2, pp. 126–131, 2016.
8. M. Zuniga and B. Krishnamachari, Analyzing the transitional region in low power wireless links, in *First Annual IEEE Communications Society Conference on Sensor and Ad Hoc Communications and Networks, IEEE SECON*, 2004.
9. K. Almi'ani, A. Viglas, and L. Libman, Tour and path planning methods for efficient data gathering using mobile elements, *International Journal of Ad Hoc and Ubiquitous Computing*, vol. 21, no. 1, pp. 11–25, 2016.
10. M.H. Eiza, T. Owens, and Q. Ni, Secure and robust multi-constrained QoS aware routing algorithm for VANETs, *IEEE Transactions on Dependable and Secure Computing*, vol.13, no. 1, pp. 32–45, 2016.
11. A. Alanazi and K. Elleithy, Real-time QoS routing protocols in wireless multimedia sensor networks: Study and analysis, *Sensors*, vol. 15, pp. 22209–22233, 2015.
12. L. Atov, H.T. Tran, and R.J. Harris, Efficient QoS partition and routing in multiservice IP networks, *Proceedings of the 2003 IEEE International Performance, Computing, and Communications*, pp. 435–441, 2003.
13. B. Deb, S. Bhatnagar, and B. Nath, Re-InForM: Reliable information forwarding using multiple paths in sensor networks, *Proceedings of the 28th Annual IEEE International Conference on in Local Computer Networks LCN '03*, pp. 406–415, 2003.
14. X. Huang and Y. Fang, Multiconstrained QoS multipath routing in wireless sensor networks, *Wireless Networks*, vol. 14, pp. 465–478, 2008.
15. A.B. Bagula and K.G. Mazandu, Energy constrained multipath routing in wireless sensor networks. In: F.E. Sandnes, Y. Zhang, C. Rong, L.T. Yang, and J. Ma (eds.) *Ubiquitous Intelligence and Computing. UIC 2008. Lecture Notes in Computer Science*, vol. 5061. Springer: Berlin, Heidelberg.
16. J.L. Lu et al., A survey on multipacket reception for wireless random access networks, *Journal of Computer Networks and Communications*, vol. 2012, p. 14, 2012.
17. G.W. Lee, S.-Y. Lee, and E.-N. Huh, Congestion prediction modeling for quality of service improvement in wireless sensor networks, *Sensors*, vol. 14, pp. 7857–7880, 2014.
18. N. Baccour et al., Radio link quality estimation in wireless sensor networks: A survey, *ACM Transactions on Sensor Networks (TOSN)*, vol. 8, no. 4, p. 34, 2012.
19. C. Plesca, V. Charvillat, and W.T. Ooi, Multimedia prefetching with optimal Markovian policies, *Journal of Network and Computer Applications*, vol. 69, pp. 40–53, 2016.
20. M. Zorzi, R.R. Rao, and L.B. Milstein, On the accuracy of a first-order Markov model for data transmission on fading channels, *Fourth IEEE International Conference on Universal Personal Communications*, vol. 1995, pp. 211–215, 1995.

21. Y. Yang et al., Network coding based reliable disjoint and braided multipath routing for sensor networks, *Journal of Network and Computer Applications*, vol. 33, pp. 422–432, 2010.
22. I.F. Akyildiz, T. Melodia, and K.R. Chowdury, Wireless multimedia sensor networks: A survey, *Wireless Communications, IEEE*, vol. 14, pp. 32–39, 2007.
23. D. Incebacak, B. Tavli, K. Bicakci, and A. Altin-Kayhan, Optimal number of routing paths in multi-path routing to minimize energy consumption in wireless sensor networks, *EURASIP Journal on Wireless Communications and Networking*, vol. 2013, p. 252, 2013.
24. F. Al-Turjman, H. Hassanein, and M. Ibnkahla, Towards prolonged lifetime for deployed WSNs in outdoor environment monitoring, *Elsevier Ad Hoc Networks Journal*, vol. 24, no. A, pp. 172–185, 2015.
25. F. Entezami and C. Politis, An analysis of routing protocol metrics in wireless mesh networks, *Journal of Communications and Networking*, vol. 4, pp. 15–36, 2014.

Chapter 4

Green Data Delivery Framework for Safety-Inspired Multimedia in Mobile IoT

4.1 Introduction

Safety services are often the most demanding in densely populated cities, suffering from various types of disasters. Some recent events with high numbers of casualties include the major earthquake that took place in 1999 in Turkey with an estimated 45,000 fatalities [1]. Snow avalanches are also serious problems, especially on highways. Furthermore, over 1.2 million people were killed in road traffic accidents around the world in 2003 according to a report published by the World Health Organization and the World Bank; another 50 million may be left injured by crashes annually. The report shows that more than 3000 people are killed in road accidents every day [1]. Thus, we focus on safety and emergency issues in this research, for which giving warnings and alerts well in advance and acting quickly in emergency situations is vital for better survivability chances.

Over the last decade, the evolution of the sensor network paradigm has introduced a well-established area of research as wireless multimedia sensor networks (WMSNs) [2]. A typical WMSN consists of a large number of multifunctional, low-cost, and low-power sensor nodes (SNs). These SNs are deployed densely and randomly in dynamically changing environments that are monitored by public cameras, and with recent advancements, they have the ability to control the multimedia content as well [2]. WMSNs perform local processing and communicate collected multimedia traffic from surrounding audio/video inputs, which have been extensively spread nowadays with the evolution of the Internet of things (IoT), to a base station that performs most of the complex processing. Thus, optimized route selection for the collected traffic is of utmost importance for timely processing in such scenarios. This can dramatically affect the energy needed to transmit the multimedia messages because it is approximately twice as great as the energy needed for text messages. Furthermore, the integration [3,4] of existing mobile wireless network (e.g., Wi-Fi and long-term evolution-advanced (LTE-A)) infrastructures in intelligence transportation system (ITS) with inexpensive WMSNs that are able to capture multimedia-related information (e.g., complementary metal-oxide semiconductor cameras) has introduced another challenge in terms of the quality of service (QoS) [2].

QoS defines a set of service requirements that need to be achieved by the WMSNs during the routing of a packet from the source to the sink. Therefore, the general research challenges for designing suitable wireless sensor routing to fulfill QoS metrics arise primarily due to the large number of these constraints that must be simultaneously satisfied. These constraints arise because of the unreliability in low-power wireless links and the limitations of availability of resources of SNs in the deployment area. Because of these limitations, there are a number of new constraints associated with supporting real-time WMSN applications [5], and it is a challenge to deliver multimedia content through typical routing protocols while maintaining high QoS. Thus, the performance of traditional SNs should be enhanced to keep up with the fast-changing realistic events in the real world.

WMSNs have been presented in several fields that require ubiquitous access in both real-time and non–real-time data. In WMSNs, SNs are mostly used to sense the continuous data and control actuators.

For example, the new emerging ITSs are demanding high quality images/videos for safety and/or security purposes. These images/videos are supposed to be exchanged in a very fast manner in between static/mobile sensors attached to cars in critical/risky situations like overtaking, as depicted in Figure 4.1. To accommodate such demands with regular WMSNs, the network designer may encounter several additional challenges that affect the data routing and delivery. For instance, network topologies are often changing, and sensors have limited power, computation,

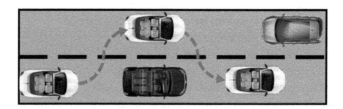

Figure 4.1 Traffic overtaking issues.

and storage capabilities. These challenges depend on the nature of the real-time multimedia data traffic, such as application-specific QoS requirements, high bandwidth, tolerable end-to-end delay, different resource constraints, multimedia source coding techniques, cross-layer coupling of functionality, and multimedia in-network processing. Therefore, to meet the QoS requirements and to use network resources in an efficient and fair manner, these challenges, along with other research design issues such as cost and connectivity, must receive attention and should be specifically considered while routing multimedia contents. Therefore, in this work, we aim at designing a cost- and energy-efficient data-delivery framework via multipath-routing/communication protocols that assures QoS for multimedia applications. Major contributions in this article can be summarized as follows:

■ A mixed-integer linear program (MILP) is used to mathematically formulate the data delivery problem in WMSNs utilized for safety-inspired ITS applications.
■ This MILP-based approach utilizes two important QoS metrics in hybrid sensor and vehicular networks (HSVNs), which are the end-to-end delay and energy consumption.
■ A reliable and real-time delivery of warning messages in HSVNs is designed to reach two main objectives, which are as follows: (1) minimizing the message delivery delay and (2) minimizing the involvement of roadside nodes in relaying tasks in order to extend the network's lifetime.

The rest of the chapter is organized as follows. Section 4.2 reviews previous related work. Section 4.3 discusses our system models. Section 4.4 describes our proposed routing approach for the IoT paradigm. Section 4.5 provides theoretical analysis studies and then the performance evaluation for the proposed approach in Section 4.6. Finally, Section 4.7 provides the conclusions and future directions.

4.2 Related Work

We focus in this work on the design of a class of data routing protocols (called multipath routing) that can fulfill the multimedia requirements. We also focus on the restrictions caused by the limited resources of the wireless sensor network (WSN). By definition, a multipath-routing operation is relevant for providing adequate network resources under varying traffic conditions in order to fulfill QoS metrics [6]. Although the strategy for a traditional single-path routing protocol can be performed by the minimization of both resource utilization and computational complexity, this might reduce the network performance because of its lower flexibility in finding the alternative paths or nodes when the discovered path fails to transmit packets to the nearest neighbor. In other words, single-path strategies do not introduce reliable solutions [7] for the routing operation. Therefore, the single-path routing strategy cannot be considered as an effective routing approach to achieve the network performance demands of various multimedia applications in the IoT era. Moreover, the application of traditional routing sensor network protocols to multimedia applications such as transmitting audio, video, and images will use up energy very quickly [8]. To overcome the limitations and the inefficiency of the single-path routing strategy, a new type of routing strategy (called multipath routing) can be used. This strategy is a promising technique in wireless networks and is thus the focus of this research.

The multipath routing strategy was first proposed by Maxemchuk [5] to spread the traffic load of incoming packets from a source to achieve load balancing and fault handling in packet-switching networks. The proposed method was demonstrated to equalize load and increase overall network performance. Since then, the multipath routing technique has been applied and utilized in the last few years for different types of network management, such as improving the delivery of multimedia content, as well as providing fault-tolerance routing and supporting QoS over large-scale Internet networks, single-hop local area networks (LANs), multihop wired LANs, automated teller machines (ATMs), broadband integrated services digital networks (B-ISDNs), and finally in communication networks such as ad hoc and sensor networks.

In general, multipath data delivery in WMSNs can be classified into unicast and multicast approaches. In the following, the state of the art of these techniques is presented, and a comprehensive analysis for the majority of multipath routing approaches is provided.

4.2.1 Multipath Unicast Forwarding

The unicast path routing protocol aims at constructing a single sequence of efficient quality links from the direct source to the sink, possibly over multiple hops, while considering the cost of a path. It is the same as the flat routing protocol operation, in which each node collaborates with the others to perform the task of constructing an optimal path that is normally called a simple flood operation. The extension of the unicast path to the multipath unicast path depends on the mechanism of construction, selection, and distribution of the optimal n-paths between the source and the sink.

4.2.1.1 Multipath QoS-Based Protocols

In general, there are several challenges associated with the design of the unicast multipath routing mechanisms for WMSNs. One such challenge is the difficulty of supporting multi-constrained QoS parameters in WMSNs because of the need of changes at various layers of the communication stack. In the following, we review a few QoS-based routing protocols. Sequential assignment routing (SAR) was the first routing protocol developed for WSNs [9] in which QoS issues were considered based on three factors: energy conservation, QoS parameters, and the level of packet-priority traffic flow. These traffic types were applied through a given flow for each data packet with a constant priority, and they remain unchanged until they reach the final destination. SAR uses a table-driven multipath approach that satisfies the QoS parameters, energy consumption, and fault tolerance. The disadvantage of SAR is that the creation mechanism of the multipath causes additional node energy depletion. Thus, it is not suitable for multimedia transmission. The stateless protocol for real-time communications in sensor networks (SPEED) is considered the first protocol to envision real-time requirements under specified QoS constraints [10]. Its localization/geographical protocol provides guarantees to support QoS parameters for real-time traffic. SPEED maintains a desired delivery speed across the network through a novel combination of the nondeterministic QoS-aware of geographic forwarding and feedback control. This combination of a medium access control (MAC) (single-hop) layer and a network (multi-hop) layer adopts a cross-layer approach that improves the end-to-end delay transmission time and provides a good response to congestion and voids. However, the disadvantage is in the prolonged lifetime of the SN that is achieved by only the reduction of control packets and geographic routing without considering other energy metrics during the routing operation. Looking deep into the multipath QoS-based protocols leads us to conclude that there is a significant need for analytical load/traffic distribution models in determining the time-space complexity (i.e., delay) of the routing methods in multimedia delivery.

4.2.1.2 Reliability Constraints

Reliability is another major challenge for the WMSNs in addition to delay and throughput. Reasons and approaches are discussed in the following. There are actually several routing protocols that have been proposed in the literature to achieve reliable data transmission in WSNs. For instance, the multipath multi-SPEED (MMSPEED) routing protocol has many improvements and modifications in order to guarantee specific QoS parameters in WSNs spanning over the network and MAC layers [11]. First, a novel packet data delivery mechanism that provides QoS differentiation in two quality domains (timeliness and reliability) is proposed. Second, an end-to-end QoS reliability according to the local decisions for each intermediate SN is proposed. This end-to-end QoS reliability is considered an important property for large-scale WMSNs.

Meanwhile, the reliable information forwarding multiple paths (Re-InForM) protocol [12] employs a probabilistic flooding to deliver the information awareness packet and service at the desired priority levels of reliability at a proportional cost for sensor networks through a dynamic and randomized multipath forwarding mechanism. The forwarding mechanism is based on the local knowledge of network conditions such as channel error, hops-count, and out-degree. Information on network conditions is stored in the header of the packet without requiring any data caching at any SN by using the dynamic packet state (DPS) method that causes an increase in the probability of information delivery. A key drawback of this protocol is the duplication of packets that might cause a high cost of energy consumption and degrade the useful channel bandwidth utilization. The network coding-reliable multipath routing (NC-RMR) protocol for WSNs was presented in [13]. The NC-RMR protocol employs computations of paths and next-hop node selection as in the Re-InForM protocol. However, it avoids the redundancy in packet copies. NC-RMR applies the network coding mechanism in delivering packets through a multipath from the source to the sink. It increases the level of reliability via employing a hop-to-hop mechanism that can establish a disjointed and braided multipath routing protocol. The main disadvantage in this protocol is that although node-disjoint multipath routing conserves energy, the selection of the path may require more hops to reach the destination.

4.2.2 Alternative Multipath Broadcasting

In this section, we discuss the alternative multipath broadcasting method in WMSNs. Broadcast flooding and its sub-taxonomy can be classified into two categories, depending on the main operation: indicator-based and indicator-free. There are different types of indicators that further categorize this type of operation into subclasses, namely a "data-centric" protocol, in which indicators are built for sensors in a setup stage. They then follow those indicators to make decisions while performing the routing operation. In the indicator-free partition, the algorithm has no initialization phase, and the packets are transmitted in an "on-demand" or random fashion.

4.2.2.1 Data-Centric Approaches

In data-centric broadcasting protocols, when a node receives a packet, it broadcasts to all neighbors until all of the nodes in the network receive the same packet through different ways based on the packet content itself. Eventually, there are significant variations among these data-centric broadcasting protocols that vary based on the targeted application. This research investigates the key categories such as the expected transmission count–delay GrEedy (EDGE), directed diffusion (DD), and energy-aware directed diffusion (EADD) [14]. If these protocols are compared, it can be easily concluded that all of them have low scalability and satisfy specific QoS requirements. For DD and EDGE, reliability is moderate, whereas it is high in EADD. Meanwhile, EDGE is moderate in terms of energy efficiency, whereas DD and EADD have high energy efficiency. However, in all of these protocols, the number of paths are unlimited, and an optimized subset of these paths should be found such that the minimum aggregation at the sink node is performed. It's also worth pointing out that the DD and EDGE protocols are partially disjoint, whereas the EADD protocol is assuming a node-disjoint path finder/routing protocol. The design of such data-centric routing protocols is usually achieved according to several aspects, such as the network functionality, data aggregation at specific nodes, location awareness, and energy efficiency. The advantages of data-centric routing are based on provisions of the nodes from individual addresses, which increases energy, aggregation, delay, and caching efficiency during data transmission through multiple paths in the network. The major disadvantage of data-centric routing is that it is usually based on a flat-topology structure that causes many problems, such as scalability, increased traffic congestion among the nodes (which are much closer to the sink), and an increase in the overhead complexity.

4.2.2.2 On-Demand Approaches

Additionally, there are routing categories (such as reactive, proactive, and hybrid protocols) that depend on how the source and the route to the final destination are discovered in an on-demand fashion. On-demand fashion protocols can be classified into two categories: (1) reactive dynamic source routing (DSR) and (2) proactive ad hoc on-demand distance vector (AODV) routing protocols. This research investigates the robust multipath DSR (RMDSR) and multipath DSR (MDSR) as examples of the reactive DSR and examines the tiny optimal node-disjoint multipath routing (TinyONDMR) protocol [15] as an example of a proactive AODV [16]. When they are compared, all of them have low scalability and high reliability. MDSR and DSR have low energy efficiency and flat network topology, whereas TinyONDMR has a hierarchical architecture. In terms of latency, RMDSR has moderate latency, and the other ones have low latency. For all of these protocols, the number of paths is unlimited, and the sink node determines the path and traffic distribution. MDSR is k-disjoint, RMDSR is partially disjoint, and the other protocols are node-disjoint. Reactive protocols are on-demand protocols that are not periodically updated and require less routing information for each node; therefore, the provision of energy and bandwidth is better than proactive routing protocols during inactivity. In real-time applications, proactive routing protocols are more appropriate because they do not require a latency in route discovery, unlike reactive routing protocols. However, there is latency in discovering the route that is called acquisition delay that may not be suitable for real-time applications. A comparison of the main multipath routing protocols that are considered in this section is shown in Table 4.1.

In this chapter, we propose two important QoS metrics in HSVNs, which are end-to-end delay and energy consumption. Road safety application imposes a timely delivery of event notifications due to its sensitive nature. In fact, the detected data by roadside nodes should be delivered in a timely manner to the sink. Furthermore, the energy consumption of deployed SNs should be minimized in order to prolong the network's lifetime. To deal with these challenges, this research estimates the optimal WMSN topology in dynamic environments via a bi-objective linear programming model with twofold aims that are minimizing the message delivery delay and decreasing energy consumption of the system. As well as introducing a novel routing protocol that aims to maximize the QoS in presence of energy-related restrictions, this work also demonstrates for the smart cities' designers the

Table 4.1 Key Design Issues and Challenges in WMSN Multipath Routing

Routing Protocol	Scalability	Reliability	Energy-Efficiency	Network Topology	Path Disjointedness	Delivery Model	QoS	Mobility
EADD [14] Shortest hierarchal routing protocol [16]	High	Restricted	Moderate	Hierarchical	Node-disjoint	Event-based	x	
SAR [9] SPEED [10] MMSPEED [11] Re-InForM [12] NC-RMR [13]	Restricted	Restricted	High	Flat	Partially disjoint	Query-based	x	x
Energy efficient [15] Hierarchal protocol [17]	Restricted	High	High	Tree	Disjoint	Query-based	x	x
2 Phase routing approach [18]	Moderate	High	Moderate	Flat	Partially disjoint	Event-based	x	x
Precision agriculture [19]	High	Restricted	High	Hierarchical	Partially disjoint	Geo-based	x	

relationship between the QoS path selection components and the way to optimize the performance of the QoS metrics in this domain. We aim to exchange interaction information to guarantee the reservation resources for the QoS requirements in multimedia sensor networks via developing an HSVN-specific QoS model. This model allows the performance of some pre-processing on the sensed/published data in order to filter and perform path-learning techniques for the future data delivery. The traffic measurements can then be used in monitoring vehicular and pedestrian traffic in smart cities under harsh/disastrous conditions.

4.3 System Model

Multi-target video monitoring [20] is emerging nowadays as one of the most important multimedia applications over the smart city paradigm [21]. In a typical multimedia setup, SNs aim at monitoring undistinguishable targets moving continuously in a given area, typically in an independent manner. To meet the monitoring demands of large-scale moving vehicles, the multimedia SNs should be able to manage/handle huge chunks of data to be routed via a multipath routing strategy toward the sink node. We propose a new architecture in which multimedia services can be supported over light sensor networks. This architecture takes into consideration numerous challenges in realizing multimedia over sensor networks such as the dynamic network topology, the varying link stats, and the limited energy budgets. In the following, we propose the intended WMSN-specific architecture for ITSs.

4.3.1 Network Architecture

The proposed HSVN system is constituted of vehicular nodes and two kinds of roadside nodes, namely SNs and roadside units (RSUs), that are deployed along the two sides of a road. The SNs are deployed with high density between two adjacent RSUs. We consider that both SNs and RSUs assume the same role, which is collecting data from the environment and transmitting it to the sink via passing vehicles that are considered as relay nodes. Furthermore, each SN in the network has the Institute of Electrical and Electronics Engineers (IEEE) 802.15.4 interface (ZigBee) dedicated to communication with roadside nodes and also with passing vehicles. Moreover, RSUs, as well as vehicular nodes, have two communication interfaces: the IEEE 802.11p interface (which is devoted to communication among vehicles through an ad hoc manner) and the ZigBee interface for communication with SNs. To achieve the objective of this HSVN, the considered system should operate as follows: when a roadside node (i.e., SN) detects an event (e.g., abnormal situations such as vehicle crashes, slippery roads, rockfalls, etc.), it proceeds firstly to send the detected information to a passing vehicle. If there is no vehicle in its transmission range, it transfers it to the nearest roadside node as illustrated in Figure 4.1.

4.3.2 Lifetime and Energy Model

The lifetime of WMSNs is defined as the time or number of transmission rounds beyond which the network can no longer deliver useful information to the outside end user. This is reflected by the network's inability to find a data delivery path with satisfactory values for quality of information (QoI) attributes such as delay, reliability, and throughput, as determined by the end user [22]. This definition not only provides information to satisfy the application requirements but also considers the status of the network and sensing resources in defining the network lifetime. It also justifies the fact that if the network does not have the necessary resources to send packets, it cannot satisfy the end user, and so it should be considered as a dead WMSN. The WMSN lifetime can therefore be evaluated based on the number of alive SNs (or RSUs). Several variants do exist with this model. The simple model identifies the time until the death of the first SN in the network as the lifetime of the network. Another variant evaluates lifetime until the death of "k" out of "n" SNs in the network, where $k < n$. The lifetime is the range between the death of "k" nodes from "n" nodes in noncritical ones [23].

In general, the network death in WMSNs can be associated with several cutoff criteria, such as the first node death, the percentage of dead nodes, or the number of dead nodes rising above a level such that the routing to the sink node is no longer possible [24], which can lead to isolated subnetworks. To mitigate this isolation problem, alternatives such as redundant nodes/links are used. This approach reduces the effect of losing some nodes due to battery exhaustion or network partitioning. Hence, network lifetime is extended. However, the network can still be considered dead if a particular percentage of dead or disconnected nodes are reached. Consequently, the network lifetime can be defined in this work as:

Definition 1

(Network Lifetime): The time span from network deployment to the instant when the percentage of alive and connected irredundant SNs/RSUs falls below a specific threshold τ.

Notice that remaining SNs and relay nodes, in addition to being alive, need to be connected to the base station (BS). For a node to be connected to the BS, it should be able to reach the BS through at least one route by using either single or multiple hops. Note also that connectivity and the percentage of living nodes are both addressed in this definition.

In order to measure the network lifetime, a measuring unit needs to be defined. In this work, we adopt the concept of a round as the lifetime metric. A round is defined as the time period over which every irredundant SN and relay node in the network communicates with the BS at least once. It can also be defined as the time span t_{round} over which each event center (EC) reports to the BS at least once. At the end of every round, the total energy consumed per node i can be written as

$$E_{cons}^i = \sum_{\text{Per round}} J_{tr} + \sum_{\text{Per round}} J_{rec}, \tag{4.1}$$

where $J_{tr} = L(\varepsilon_1 + \varepsilon_2 d^n)$ is the energy consumed for transmitting a data packet of length L to a receiver located d meters from the transmitter. Similarly, $J_{rc} = L\beta$ is the energy consumed for receiving a packet of the same length. In addition, ε_1, ε_1, and β are hardware-specific parameters of the used transceivers. In addition, if the initial energy E_{init} of each node is known, the remaining energy per node i at the end of each round is $E_{rem}^i = E_{init}^i - E_{cons}^i$.

4.3.3 Communication Model

Radio interference, antenna shape and orientation, distance, and environmental factors may vary during the network lifetime and affect link quality between the SNs. Despite the fact that the locations of SNs are fixed, as well as every node being configured with the same transmission range, environmental variations result in asymmetric links between nodes. We assume a probabilistic model in which the probability of communication between two wireless devices decays exponentially with distance and takes into consideration surrounding obstacles and hindrances. Accordingly, the communication range of each device can be represented by an arbitrary shape. For realistic estimation of the arbitrary shape dimensions, we need a practical signal propagation model. This model can describe the path loss* in the targeted site by taking into consideration the effects of the surrounding terrain on the power (P_r) of received signals as follows [25]:

$$P_r = K_0 - 10\gamma \log(d) - \mu d, \tag{4.2}$$

where d is the Euclidian distance between the transmitter and receiver, γ is the path loss exponent calculated based on experimental data, μ is a random variable describing signal attenuation effects[†] in the monitored site, and K_0 is a constant calculated based on the transmitter, receiver, and field mean heights. Let P_r equal the minimal acceptable signal level to maintain connectivity. Assume γ and K_0 in Equation 4.2 are also known for the specific site to be monitored. Thus, a probabilistic communication model that gives the probability that two devices separated by distance d can communicate with each other is given by

$$P_c(d,\mu) = Ke^{-\mu d^\gamma}, \tag{4.3}$$

where $K_0 = 10 \log(K)$. Thus, the probabilistic connectivity P_c is not only a function of the distance separating the SNs but is also a function of the surrounding obstacles and terrain, which can cause shadowing and multipath effects (represented by the random variable μ).

---

* Path loss is the difference between transmitted and received signal power.
[†] Wireless signals are attenuated because of shadowing and multipath effects. This refers to the fluctuation of the average received power.

4.4 Multipath Disruption-Tolerant Approach (MDTA)

This approach utilizes two important QoS metrics in hybrid sensor and vehicular networks, which are end-to-end delay and energy consumption. Road safety application imposes a timely delivery of event notifications due to its sensitive nature. To model the problem of reliable and real-time delivery of warning messages in HSVNs, this approach is designed to reach two main objectives, which are as follows: (1) minimizing the message delivery delay and (2) minimizing the involvement of roadside nodes in relaying tasks in order to extend the network's lifetime.

Assume F_r: the set of the roadside nodes (SNs and RSUs) deployed along the two sides of the road; F_v: the set of vehicles traveling on the road; F: the set of all nodes in the network except the sink, $F = F_r \cup F_v$; F_{Info}: the set of all nodes in F, such that every node holds information that can be received from a neighbor or detected by itself; M: the total number of pieces of information or events which can occur on the road during a fixed period; T_{ij}^k: the time needed to transfer the piece of information k from node i node j; R_i: the transmission range of node i, such that $R_i = R_S$ if i is a SN, $R_i = R_{SU}$ if i is an RSU, and $R_i = R_V$ if i is a vehicle; (a_i, b_i): the coordinates of node i; $d(i,j)$: the Euclidean distance between two nodes i and j, $d(i, j) = \sqrt{(a_i - a_j)^2 + (b_i - b_j)^2}$; $V(i)$: the set of nodes belonging to $F \cup \{Sink\}$, which are within the transmission range of node i; formally $V(i) = \{j \in F \cup \{sink\}: d(i, j) \leq R_i\}$. Thus, the targeted problem can be formulated as a bi-objective optimization program as follows:

$$\min \sum_{j \in F} \sum_{i \in V(j)} \sum_{k=1}^{M} T_{ij}^k y_{ij}^{kd_j} \tag{4.4}$$

$$\min \sum_{j \in Fr} \sum_{i \in V(j)} \sum_{k=1}^{M} y_{ij}^{kd_j} \tag{4.5}$$

subject to

$$\sum_{j \in V(j)} \sum_{k=1}^{M} y_{ij}^{kd_j} = 1, \quad \forall i \in F_{Info} \tag{4.6}$$

$$\sum_{k=1}^{M} y_{ij}^{kd_j} \geq \sum_{k=1}^{M} y_{it}^{kd_t}, \quad \forall F - \{Sink\}, \quad \forall j, t \in V(i) \text{ with } d_j < d_t \tag{4.7}$$

$$\sum_{k=1}^{M} y_{isink}^{k0} = 1, \quad \forall i \in V(Sink) \cap F_{Info} \tag{4.8}$$

$$y_{sinkj}^{kd_j} = 0, \quad \forall j \in V(Sink), \quad \forall k = 1,2,3,\dots M \tag{4.9}$$

$$y_{sinkj}^{kd_j} \in \{0,1\}, \quad \forall i, j \in F \cup (Sink), \quad \forall = 1,2,3,\dots M \tag{4.10}$$

$$y_{ij}^{kd_i} = \begin{cases} 1, & \text{if the node } j \text{ located at a distance } d_j \text{ from sink} \\ & \text{receives the information } k \text{ sent by the node } i, \\ 0, & \text{otherwise.} \end{cases} \tag{4.11}$$

The aim of the first objective Equation 4.4 is to minimize the sum of end-to-end delay of all sensed pieces of information (from the source node until its successful reception by the sink) corresponding to the events that have occurred in the road during a fixed period. The second objective function given by the Equation 4.5 consists of minimizing the participation of roadside nodes in relaying tasks by minimizing the two types of communication, namely "roadside node to roadside node" and "vehicle to roadside node." Thus, the objective is to incite the nodes to send the detected or received data as much as possible to the passing vehicles instead of to the roadside nodes. The first constraint 4.6 indicates that once a roadside node receives or detects any piece of information k, it has to forward it to only one neighbor (assuming that at each instant a roadside node cannot hold more than one piece of information). This limitation of the generated transmissions allows the SNs to preserve their restricted energy and network overhead. The constraint 4.7 guarantees that if a node (either a roadside node or a vehicle) holds any piece of information, it will forward it to whichever of its neighbors is nearest to the sink. The constraint 4.8 assures that any node in the neighborhood of the sink which holds any piece of information has to forward it directly to the sink. This ensures that the data transmission ends at the sink and the sensed information will certainly reach it. The constraints 4.9 through 4.11 signify that once the sink receives a piece of information, it does not send it back to any other node in its transmission range. The proposed approach is aimed to ensure an efficient and timely delivery of warning messages from a source node (i.e., the node that detects any anomaly around the road) until the sink. In other words, the generated information at one SN must actually be routed in a timely fashion to the sink. Toward this end, this strategy is aimed at minimizing the total time taken by the detected piece of information from a source node until it reaches the sink. In fact, this approach deals with the restricted energy capacities of SNs to extend the network's lifetime by minimizing the involvement of the deployed roadside nodes, which is possible by inciting them to essentially perform sensing tasks and letting the relaying data function to the oncoming vehicles as much as possible. In other words, the roadside nodes act as routers to forward the sensed piece of information to vehicles or the sink just in necessary circumstances, such as instances of low vehicle density or network disconnections. When these special circumstances do not occur, the SNs remain in an "inactive" state

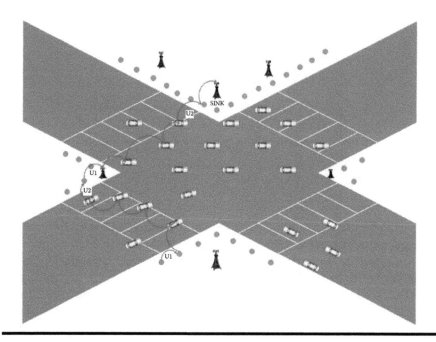

Figure 4.2   The considered scenario in an HSVN.

(performing neither reception nor transmission); this allows the SNs to conserve their energy and extend the network's lifetime. Moreover, when sensed data are received by a vehicle, it is forwarded from one vehicle to another via vehicle-to-vehicle (V2V) communication until it reaches the sink. If in some cases a vehicle does not have another vehicle in its transmission range, which means that the vehicular network is in a disconnection situation, it sends back the obtained information to the nearest roadside node, as depicted by the arcs labeled U2 in Figure 4.2. The detailed description of the proposed MDTA can be summarized in Algorithm 4.1.

Algorithm 4.1: Pseudo-code of the MDTA

Function: Path selection in MDTA

1. F: all nodes in the network except the sink.
   NV_x: Vehicles that are neighbors of the node x in the network.
   NR_y: Roadside nodes (SNs or RSUs) that are neighbors of the node y in the network.
2. for $s = 1$ to size (F) do
3. if s holds a piece of information k then
4. build NVs;
5. end if

6. while k has not reached the sink do
7. if NVs = Ø then
8. /*Select next_hope*/
9. for $i = 1$ to size (NVs) do
10. calculate the distance (i, sink);
11. end for
12. next_hope ← the nearest vehicle to the sink;
13. send to the next_hope;
14. else
15. build NRs;
16. for $i = 1$ to size (NRs) do
17. calculate the distance (i, sink);
18. end for
19. next_hope ← the nearest roadside node to the sink;
20. if next_hope's state = active then
21. send to the next_hope;
22. else
23. activate (next_hope);
24. send_to (next_hope);
25. end if
26. end if
27. end while
28. end for

In the above algorithm, each node that holds a piece of information builds a set of all its neighbors (steps 2–5). First, it checks if there are nodes among them that are vehicles, and it computes the distance of each one to the sink (steps 6–11). Then, the holding piece of information will be sent to the nearest vehicle to the sink (steps 12 and 13). If there is no vehicle in its transmission range (vehicular ad hoc network [VANET] disconnection), it activates the nearest neighbor in the wireless sensor network, which is selected as a next hop (steps 15–25). This process is repeated until the piece of information successfully reaches the sink.

4.5 Theoretical Analysis on Lifetime

In the previous subsection, we examined the placement problem for a heterogeneous WMSN when both energy-efficient and delay-tolerant design factors are considered. However, once the network becomes operational, deployed nodes start losing energy and face harsh operational conditions that might lead to failures and separations. Because these conditions are scarcely predictable in practice, it is very difficult to predict what would be the maximum number of rounds for which a WMSN can stay operational. Therefore, we derive an upper bound on the number of rounds for which a WMSN can stay operational, given that there are no unexpected node/link failures. Also, we define LT_{max} to be the maximum number of rounds for which a WMSN can stay operational, $E_{min/r}^{SN}$ to be the minimum total energy consumed by SNs nodes per round, and $E_{min/r}^{RN}$ to be the minimum total energy consumed by relay node per round. Assume E_{init}^{tot} is the initial total available energy before the network becomes operational. Assume E_{init}^{SN} is the initial available energy per SN. Assume E_{init}^{RN} is the initial available energy per relay node. N_{SN} is the total number of SNs. N_{RN} is the total number of relay nodes.

Theorem 4.1

An upper bound on the deployed network lifetime is

$$LT_{max} = \min\left\{ \frac{N_{SN} \cdot E_{init}^{SN}}{J_{tr}\left\lceil \frac{N_{SN}}{k_1} \sum_{i=1}^{N_{SN}/k_1} G_i \right\rceil}, \frac{N_{RN} \cdot E_{init}^{RN}}{(J_{tr} + J_{rc})\left\lceil \frac{N_{RN}}{k_2} \sum_{i=1}^{N_{RN}/k_2} RG_i \right\rceil} \right\}. \quad (4.12)$$

Proof. Because the minimum consumed energy per round by SNs is the required energy to deliver irredundant generated traffic (sensed data), the minimum energy consumed by these sensors per round is equal to the energy used in transmitting from irredundant SNs. Since the irredundant SNs are equal to N_{SN}/k_1, the minimum energy consumed per round is

$$E_{min/r}^{SN} = J_{tr}\left\lceil \frac{N_{SN}}{k_1} \sum_{i=1}^{N_{SN}/k_1} G_i \right\rceil. \quad (4.13)$$

Because the number of irredundant relay nodes is equal to N_{RN}/k_2, the minimum energy consumed per round is

$$E_{min/r}^{RN} = (J_{tr} + J_{rc})\left\lceil \frac{N_{RN}}{k_2} \sum_{i=1}^{N_{RN}/k_2} RG_i \right\rceil. \quad (4.14)$$

Because the initial total available energy at SNs is equal to $N_{SN} \cdot E_{init}^{SN}$, the maximum number of rounds for which the SNs can stay operational is $N_{SN} \cdot E_{init}^{SN}/E_{min/r}^{SN}$. Similarly, the maximum number of rounds for which the relay node can stay operational is $N_{RN} \cdot E_{init}^{RN}/E_{min/r}^{RN}$. Because the maximum number of rounds for which a WSN can stay operational is controlled by the lifetime of the SNs generating the sensed data and the relay nodes relaying this data, the maximum number of rounds for which a WSN can stay operational is

$$LT_{max} = \min\left\{\frac{N_{SN} \cdot E_{init}^{SN}}{E_{min/r}^{SN}}, \frac{N_{RN} \cdot E_{init}^{RN}}{E_{min/r}^{RN}}\right\}. \qquad (4.15)$$

By substituting Equations 4.13 and 4.14 in 4.15, we achieve the lifetime upper bound in Equation 4.12 above.

It is worth pointing out that this upper bound depends only on the number of deployed nodes, initial node energy, node generation rate, redundancy level (i.e., k-value), and energy consumed per packet transmitting/receiving. Further improvements to the bound are possible by increasing the initial energy of the deployed nodes, by decreasing their energy consumption per packet, by increasing the number of nodes, or by increasing the redundancy level. This upper bound is used not only in assessing the efficiency of our proposed MDTA approach but also in any other routing strategy aimed at maximizing the WMSN lifetime.

4.6 Performance Evaluation

In this section, we evaluate the performance of the proposed MDTA. We use SPEED, MMSPEED, EADD, and MDSR as baseline evaluation algorithms. These algorithms are applied on 10 randomly generated WMSN mesh topologies in order to get statistically stable results. For each topology, we apply a random node/link failure, and performance metrics are computed accordingly. We make use of OMNeT++ and incorporate the use of MOVE for modeling traffic and mobile resources. We assume a predefined random time schedule for traffic generation toward the sink.

4.6.1 Performance Metrics and Parameters

To compare the performance of these five schemes, the following three performance metrics are used.

1. Average delay: This is measured in msec and is defined as the average amount of time required to deliver a data unit to the destination.
2. Idle time: This metric reflects the ratio of idle time every node spends while waiting to forward a message. It is measured in μsec.
3. Throughput: This is set here as a quality measure. It is the average percentage of transmitted data packets that succeed in reaching the destination reflecting the effect of node heterogeneity and delay in IoT setups over the utilized data delivery approach.

Meanwhile, the three data delivery performance metrics are assessed by using the following four parameters:

1. The size of the network in terms of total node count. This reflects the application's complexity and the scalability of the exploited routing scheme. It should be noted that a larger node count in a data path raises the risk of node failure and, hence, of dropped packets.
2. Average energy consumption rate per data unit (π'_ι) as an indicator of the network power saving. This metric is measured in kJ.
3. Probability of node failure (PNF [%]): The probability of physical damage and/or battery depletion for the deployed SN due to outdoor harsh operational conditions. This parameter is chosen to reflect the impact in case of undesired circumstances under hazardous and emergency conditions and/or fragmented networks in the IoT.
4. Average cost (γ_i): This metric is used to observe the influence of the utilized data delivery approach on the overall cost to deliver a data unit from source to destination on average. The cost charged by each node n_i as γ_i is

$$\gamma_i = c_i * \left[\frac{E_{Tx}(D_k, n_j) + E_{Rx}(D_k)}{e_i} + \pi'_\iota + \acute{u}_\iota \right], \qquad (4.16)$$

where $\acute{u}_\iota$ is the available buffer space at node i, and π'_ι is the power amount to be consumed per packet processing at node i. $E_{Tx}(D_k, n_j)$ and $E_{Rx}(D_k)$ are the amounts of energy used to transmit a data packet D_k from node i to j and receive a data packet D_k at node i, respectively. And e_i is the instantaneous available energy per node i. c_i is a pricing factor for each node in the IoT measured in $ per byte. This is a factor that could be set as a flat rate per number of bytes transmitted or computed based on the state of the current resources at node n_j, in which setting it to a relatively high value would diminish the chances of n_i to be selected for relaying the data packet D_k.

4.6.2 Simulation Setup

We assume up to 1500 total SNs communicate with one sink via 100 vehicles. All mobile nodes (vehicles) are set to follow the random waypoint mobility model [16]. We used OMNeT++ as a simulation tool for this purpose. It determines whether a wireless node is connected to its neighbors or not based on the aforementioned probabilistic communication model, where $P_c = 70\%$.

The simulation is processed in three platforms, which are Windows, Linux, and OSX for validation purposes. We executed our simulation 10 times for each experiment and plotted the average results. The average results hold a 95% confidence interval no more than 5% of the average (over 500 runs). More details about our simulation are shown in Table 4.2.

Table 4.2   Simulation Parameters and Values

Parameter	Value
Targeted area	1000 m × 1000 m
Number of nodes	SNs: 350, vehicles: 100, sink: 1
Communication range	SN: 142 m, vehicle: 300 m, sink: 500 m
Initial energy	SN: 31 kJ, vehicle: 110 kJ, sink: *Unlimited*
Energy consumption	SN/vehicle (receiving): 31.2 uJ/bit SN/vehicle (transmitting): 53.8 uJ/bit
Current consumption in sleep mode: I_{sleep}	1 μA
Current consumption in receive mode: I_{rx}	20 mA
Current consumption in active mode: I_{ac}	100 mA
Current consumption while transmitting	120 mA
Traffic intensity	90%
Log-normal shadowing variance	0, 2 dB, 4 dB or 6 dB
Bit error rate (BER) required (QoS)	10^{-4}
Radio frequency (RF) bandwidth used	200 kHz

4.6.3 Simulation Results

In the following figures, the number of SNs (i.e., RSUs) is fixed to 350 to see the effect of vehicular relaying nodes and PNF = 0.2. In Figure 4.3, the experienced delay in delivering a data packet is plotted against the size of the network for the different simulated algorithms. We observe that SPEED has the highest delay, whereas MDTA has the lowest delay as the number of nodes increases. Therefore, we can say that MDTA is more delay tolerant in comparison to all of the sampled algorithms. We also observe that there is a monotonic increase in delay for the SPEED algorithm, whereas MDSR has a slightly higher delay than MDTA, with a constant difference at every node. For MDTA and MDSR, we observe a steep increase between 100 and 200 nodes, whereas SPEED has its steepest slope between 150 and 200. For EADD and MMSPEED, we observe a fairly continuous increase in delay as the number of nodes increases because they are more dependent on the network nodes' geolocations.

Figure 4.4 shows the experienced network throughput versus the number of nodes for the sampled algorithms. We can observe that there is a general increase in

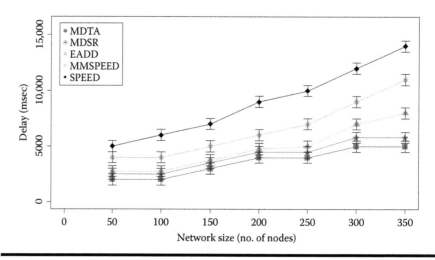

Figure 4.3   Delay versus the number of nodes in an IoT network.

throughput of the sampled algorithms as the size of the network increases. MDSR, EADD, and MMSPEED have the same throughput until the size of the network is about 150 nodes, after which MDSR gives a higher throughput. Also, it's worth remarking here that EADD adds redundant packets in order to increase the packet delivery probability while experiencing link error periods. This leads to a significant increase in the overall throughput in comparison to MMSPEED and SPEED methods. From the graph, we can also observe that in all instances, MDTA has a higher throughput than the others do, and SPEED has the lowest throughput. Hence, we can conclude that MDTA has a better throughput as the network size increases

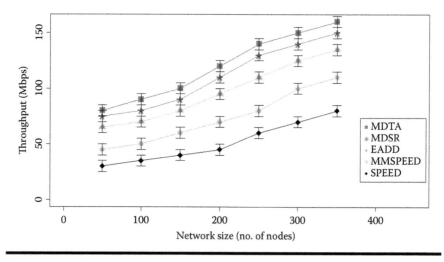

Figure 4.4   Throughput versus the network size.

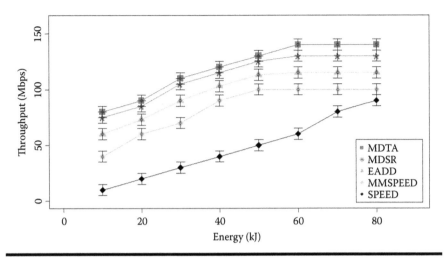

Figure 4.5 Throughput versus the available energy budget.

compared with the sampled algorithms. This can be returned to the efficient retransmission approach in the MDTA algorithm in comparison with other approaches in the literature. This makes it the most scalable approach for the next generation of IoT networks in which the connected network nodes are dramatically increasing day after day.

Plotted curves in Figure 4.5 show the average consumed energy against throughput for the different examined algorithms. We notice that there is almost a linear increase in energy consumption while applying the SPEED approach as the network throughput increases, whereas MDTA, MMSPEED, EADD, and MDSR form concave-like curves. We also observe that for every amount of energy consumed, SPEED has the lowest throughput. On the other hand, MDTA has the highest throughput for the same amount of energy. For this reason, we can conclude that MDTA is the most efficient algorithm in terms of energy consumption compared to the sampled ones. Moreover, we notice that when the energy budget is greater than or equal to 60 kJ, the network throughput is saturated due to other design factors such as network size and cost factor (c_i).

Figure 4.6 shows the average charged cost (c_i) per network node i against the experienced data delivery delay for all the simulated approaches. From this figure, we can observe that with the increase in c_i, there is a general decrease in delay time for all the sampled schemes. This is an expected network behavior because the flat rate charge increases per node. We also notice that all the schemes reach a certain threshold, at which point the delay becomes constant at 2000 milliseconds. MDTA is the first to get to the threshold when c_i equals 40, whereas SPEED is the last when c_i equals 60. MMSPEED and MDSR reach the threshold c_i equals 50. Consequently, we can say that MDTA is the most cost-effective scheme since it has the lowest delay time with the lowest c_i.

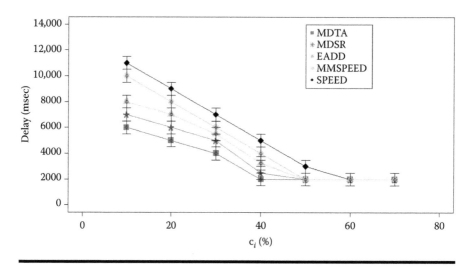

Figure 4.6   Delay versus the average pricing rate percentage c_i.

In Figure 4.7, the average idle time is compared under the varying total count of network nodes. As the network size or the number of SNs increases, there is a general increase in the average idle time. However, we observe that MDTA has the lowest idle time compared with other baselines. From Figure 4.7, we can also deduce that after a network size of 150 nodes, the average idle time of MDTA remains constant, which means it is not affected by the number of nodes. MDSR has a slightly higher idle time than MDTA. MDSR's difference to MDTA remains constant as the network increases in size. MMSPEED and SPEED have an increasing

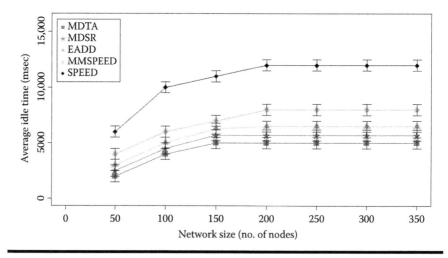

Figure 4.7   Average idle time versus the network size.

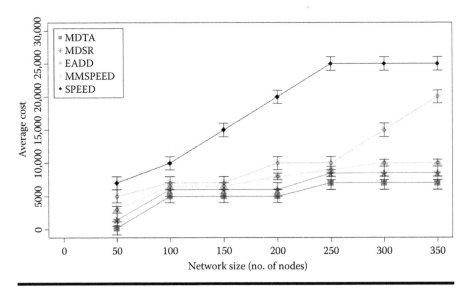

Figure 4.8   Average cost versus the network size.

idle time until 200 nodes, at which point they stay in a steady state. Therefore, we can conclude that MDTA is the most efficient compared with the sampled baseline algorithms.

Figure 4.8 depicts the network size against the average cost (γ_i) of all the schemes. From the figure, we can observe that SPEED has the highest average cost. On the other hand, the MDTA approach has the lowest average cost under all varying node counts. When the number of nodes reaches 250, the SPEED approach has a constant and fixed average cost. On the contrary, after a network size of 250 nodes, we observe a sharp and linear increase on MMSPEED and EADD. Meanwhile, MDSR is the scheme that has the second-lowest average cost after the MDTA approach. The achieved cost curve of MDSR closely follows that of MDTA. However, it is still worse than MDTA. Therefore, MDTA has the best performance in terms of average cost, as well as having the best performance under all experimented network sizes. The reason is that the MMSPEED, EADD, and MDSR approaches add redundant packets in order to increase the packet delivery probability while experiencing link error periods. This leads to significant increases in the overall cost.

Furthermore, when we compare these algorithms in terms of the number of transmission rounds, it can be clearly observed from the simulation results in Figure 4.9, that MDTA outperforms MMSPEED, EADD, and MDSR. Notably, more savings in terms of the amount of remaining energy achieved via the proposed routing approach MDTA caused by applying energy constraints have led to prolonged network lifetime, as shown in Figure 4.9.

Moreover, we examined the four routing approaches (MDTA, EADD, MMSPEED, and MDSR) in terms of the average delay impacts (Figure 4.10)

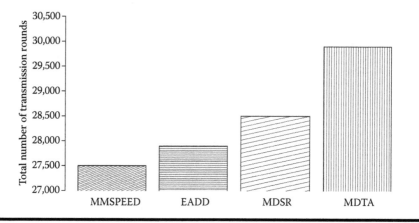

Figure 4.9 Comparison of the three data delivery techniques based on total number of transmissions.

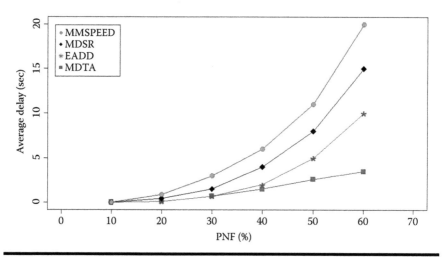

Figure 4.10 Average delay versus the probability of node failure in the network.

while considering disaster scenarios and/or fragmented networks, in which the failure of a critical node partitions the network into disjoint segments. Based on Figure 4.10, we notice a severe effect on the average delay while the PNF is increasing. We notice that all approaches are experiencing an exponential increase in the experienced delay as the network becomes disconnected. However, by using the proposed MDTA approach, the increase is going linear, which can be a very desirable feature in the IoT while experiencing harsh operational conditions and severe mobility effects.

4.7 Conclusions

In this chapter, we investigated WMSN routing techniques for the IoT paradigm in terms of energy consumption, cost, and delay while experiencing harsh operational conditions and severe energy limitations. We proposed a novel approach for sensor networks in the IoT called MDTA. We found that MDTA can save a considerable amount of energy. Moreover, we showed how the data delivery price can be affected by the network size for varying energy-based routing approaches. This approach uses a reliable communication protocol between VANETs and WSNs for an efficient traffic road information exchange among the different nodes. For this reason, it is aimed at finding the best path between a source node and the sink in terms of end-to-end delay and energy consumption. Thus, the model has two objectives: ensuring timely delivery of warning messages detected by the deployed SNs and dealing with the challenge of extending the network's lifetime by minimizing the participation of SNs in the relaying function in vehicular environments. As for fulfilling future prospects, it would be interesting to deal with large-scale routing problems considering a large hybrid sensor and vehicular networks and to introduce some additional features for routing problems in vehicular networks, such as location and the speed of vehicles.

References

1. V.I. Marza, On the death toll of the 1999 Izmit (Turkey) major earthquake, *ESC General Assembly Papers*, European Seismological Commission: Potsdam, 2004.
2. I.F. Akyildiz, T. Melodia, and K.R. Chowdury, Wireless multimedia sensor networks: A survey, *Wireless Communications*, vol. 14, no. 6, pp. 32–39, 2007.
3. L. Bölöni and D. Turgut, Value of information based scheduling of cloud computing resources, *Future Generation Computer Systems Journal (Elsevier)*, 2017, http://dx.doi.org/10.1016/j.future.2016.10.024
4. N. Kumar, R. Iqbal, S. Mistra, and J. Rodrigues, Bayesian coalition game for contention aware reliable data forwarding in vehicular mobile cloud, *Elsevier Future Generation Computer Systems*, vol. 48, no. 1, pp. 60–72, 2014.
5. N. Pantazis et al., Energy-efficient routing protocols in wireless sensor networks: A survey, *Communications Surveys & Tutorials, IEEE*, vol. 3, pp. 1–41, 2012.
6. S. Ehsan and B. Hamdaoui, A survey on energy-efficient routing techniques with QoS assurances for wireless multimedia sensor networks, *Communications Surveys & Tutorials*, vol. 14, pp. 265–278, 2012.
7. D. Turgut and L. Bölöni, Value of information and cost of privacy in the Internet of things, *IEEE Communications Magazine*, vol. 55, no. 9, pp. 62–66, 2017.
8. L. Yanjun et al., Enhancing real-time delivery in wireless sensor networks with two-hop information, *IEEE Transactions on Industrial Informatics*, vol. 5, pp. 113–122, 2009.
9. T. Vairam and C. Kalaiarasan, Interference aware multi-path routing in wireless sensor networks, *International Journal of Emerging Science and Engineering (TM)*, vol. 1, no. 10, pp. 74–78, 2013.

10. H. Tian et al., SPEED: A stateless protocol for real-time communication in sensor networks, in *Proceedings 23rd International Conference on Distributed Computing Systems*, 2003, pp. 46–55.

11. E. Felemban et al., MMSPEED: Multipath multi-SPEED protocol for QoS guarantee of reliability and. Timeliness in wireless sensor networks, *Mobile Computing, IEEE Transactions on*, vol. 5, pp. 738–754, 2006.

12. B. Deb et al., ReInForM: Reliable information forwarding using multiple paths in sensor networks, in *Proc. IEEE International Conference on in Local Computer Networks*, 2003, pp. 406–415.

13. Y. Yang et al., Network coding based reliable disjoint and braided multipath routing for sensor networks, *Journal of Network and Computer Applications*, vol. 33, pp. 422–432, 2010.

14. C. Jisul and K. Keecheon, EADD: Energy aware directed diffusion for wireless sensor networks, in *International Symposium on Parallel and Distributed Processing with Applications*, ISPA '08, 2008, pp. 779–783.

15. J. Ben-Othman and B. Yahya, Energy efficient and QoS based routing protocol for wireless sensor networks, *Journal of Parallel and Distributed Computing*, vol. 70, pp. 849–857, 2010.

16. C.J. Barenco Abbas et al., SHRP: A new routing protocol to wireless sensor networks, in *Advances in Computer Science and Engineering*, vol. 6, H. Sarbazi-Azad et al., Eds., Springer: Berlin, pp. 138–146, 2009.

17. E. Gelenbea et al., Routing of high-priority packets in wireless sensor networks, *IEEE Communications*, vol. 66, 2008.

18. F. Al-Turjman, H. Hassanein, and M. Ibnkahla, Optimized relay repositioning for wireless sensor networks applied in environmental applications, in *Proceedings of the International Wireless Communications and Mobile Computing Conference*, Istanbul, Turkey, 2011, pp. 1860–1864.

19. M. Biglarbegian and F. Al-Turjman, Path planning for data collectors in precision agriculture WSNs, in *Proceedings of the International Wireless Communications and Mobile Computing Conference (IWCMC)*, Nicosia, Cyprus, 2014, pp. 483–487.

20. H. Dai, Z. Zhu, and X. Gu, Multi-target indoor localization and tracking on video monitoring system in a wireless sensor network, *Journal of Networks and Computer Applications*, vol. 36, no. 1, 2013.

21. A. Aburumman and K.K.R. Choo, A domain-based multi-cluster SIP solution for mobile ad hoc network, in *Proceedings of the International Conference on Security and Privacy in Communication Systems*, Beijing, China, 2014, pp. 267–281.

22. F. Al-Turjman, H. Hassanein, and M. Ibnkahla, Towards prolonged lifetime for deployed WSNs in outdoor environment monitoring, *Ad Hoc Networks*, vol. 24, no. A, pp. 172–185, 2015.

23. F. Al-Turjman and H. Hassanein, Enhanced data delivery framework for dynamic Information-Centric Networks (ICNs), in *Proceedings of the IEEE Local Computer Networks (LCN)*, Sydney, Australia, 2013, pp. 831–838.

24. H. Fang, L. Xu, and K.-K.R. Choo, Stackelberg game based relay selection for physical layer security and energy efficiency enhancement in cognitive radio networks, *Elsevier Applied Mathematics and Computation*, vol. 296, no. 1, pp. 153–167, 2017.

25. J. Rodrigues, S. Fraiha, H. Gomes, G. Cavalcante, A. de Freitas, and G. de Carvalho, Channel propagation model for mobile network project in densely arboreous environments, *Journal of Microwaves and Optoelectronics*, vol. 6, no. 1, pp. 236–248, 2007.

Chapter 5

A Delay-Tolerant Framework for Integrated RSNs in IoT

5.1 Introduction

The Internet of things (IoT) has gained rapid attention as a comprehensive paradigm that is driven by an expansion of the Internet in which every device and physical object is uniquely identified and accessible [1]. Technological developments in the fields of item identification and wireless sensor networks (WSNs) define the characteristics of this ultra-large topology in which things are tagged and sensed instantaneously [1–3]. The IoT embodies all forms of wireless telecommunication and directly influences supply chain management, transportation, health care, and disaster alerting, among many other disciplines. Hence, the IoT is considered to be the most ambitious and disruptive phase of the Internet revolution [4], with a reach that includes hundreds of billions of objects [5,6] as well as living entities [7] that are augmented with sensing, processing, and networking capabilities. This implies significant design requirements and challenges stemming from the levels of heterogeneity and desired connectivity among the objects involved.

Since a wide portfolio of devices, services, and networks will eventually build up the IoT, the integration of numerous wireless technologies is inevitable in order to overcome design and interoperability challenges. One of the forefront technologies driving the IoT vision is radio-frequency identification (RFID) [8]. RFID

153

systems, consisting basically of readers and tags, are favored because of their non-disruptive small size and low cost and because their lifetimes are not limited by battery duration. However, RFID systems on their own do not suffice in delivering the IoT vision since other emerging applications require reporting information beyond the mere identity or location of an object, especially as context-aware services are expanding as IoT applications [9]. For the most part, such applications rely on WSNs for sensing and monitoring. In fact, WSN technology is directly contributing to the development of the IoT [2–4].

The integration of RFIDs and WSNs will increase their combined data reporting capabilities. RFIDs will provide WSNs with a limitless pool of identities (IDs) that is more reliable and scalable than the traditional medium access control (MAC) addressing. WSNs, on the other hand, being able to monitor physical events, can provide much more information on the measurement of temperature, humidity, pressure, etc., than simple RFID. Moreover, in an RFID system, reader-tag communications are conducted in single hops without intercommunication among tags. RFID systems integrated with WSNs will enjoy the advantages of wireless multi-hop intercommunication over wider areas. Additional common WSN features include cooperative applications and events triggered inside the network. These are characteristics of active networking that are required by clustering designs [10] for the IoT to create collaborative, multi-hop, and dynamic interactions among objects equipped with wireless embedded devices.

For all the aforementioned reasons, it is then only natural to consider the integration of WSNs and RFID tags into RFID-sensor networks (RSNs). Such integration is a promising approach that will result in more functional, scalable, and cost-effective systems. It would also enable a plethora of new applications in the IoT framework through supporting the sensing, computing, and communication capabilities into an otherwise passive system. However, these technologies were built on different design requirements. Although they share the utility of wireless mediums, the design goals vary in many aspects. Thus, their integration faces many challenges and constraints. The first concern relates to resource consolidation and utilizing different communication link capacities that vary in their rates of throughput. There is also the cost constraint related to designing and deploying complex integrated hardware components. These components are more complicated than simple sensors or tags. Thus, nonredundant deployment of such devices is a critical factor in determining the cost-efficiency of any integrated RSN system.

In addition to packet loss and cost-efficiency, node mobility introduces another design challenge. Mobility causes frequent partitioning of the network. Subsequently, the IoT nodes are consistently unaware of the existence of their neighbors and lack the knowledge of the future topology of the network. Thus, a contemporaneous end-to-end path between any pair of nodes is not guaranteed in such a setting. Advances in disruption/delay-tolerant networks (DTNs) [11] cater for excessive periods of disruption, in which concurrent end-to-end links may not exist. Hence, nodes need to buffer the messages until a suitable forwarding opportunity appears, or even better, until the message's assigned destination is encountered. This is known in DTN literature as store-carry-forward (SCF) routing [12].

In this chapter, we introduce a framework for node deployment and delay-tolerance in RSNs under the IoT paradigm. Our framework comprises two components. The first is SIWR,* a novel, smart, integrated WSN and RFID architecture that classifies nodes into light nodes (LNs) and integrated super nodes. The architecture is based on designing super nodes to perform the tasks of RFID-reading and wireless relaying, simultaneously, in order to dedicate the battery power of LNs to the sole task of data collection and thus prolong the lifetime of the system as a whole. The cost-efficiency of that architecture stemmed from developing a function to minimize the count of costly super nodes in the topology. This is achieved by providing a novel integer linear programming (ILP) formulation for optimal placement of super nodes while fulfilling connectivity constraints.

However, integrated IoT architectures face significant connectivity challenges. The intermittent nature of connections over IoT layouts discussed earlier presents delay either as a constraint or an objective function to evaluate RSN routing approaches. Delay could be considered as a factor that requires optimization over the whole network. To this end, in the second part of our framework, we introduce an optimized delay-tolerant approach for integrated RSNs (DIRSN).

DIRSN is a novel scheme for data routing and courier nodes' (CNs') selection in RSNs. It takes into account that super nodes may lose the connection to their corresponding access points (APs). The DIRSN approach considers the variations between nodes in an IoT layout in terms of mobility and connectivity capacities. By associating these variations with the fact that an IoT layout is frequently disrupted, we introduce a new category of nodes to our architecture, CNs, and employ them into a new decentralized ILP-based delay-tolerant approach that locates the optimum set of CNs per time round. This approach aims toward guaranteeing minimum-delay connectivity between super nodes and APs by locating the best set of CNs capable of doing so without violating the main requirements of RFIDs and WSNs that are imposed by the dense deployment and node/link capacities, such as load balancing.

Our framework aims toward providing an optimized architecture for integrated RSNs in addition to a delay-tolerant routing scheme. Although there have been many attempts to introduce integrated RSN architectures [14–20], there does not exist in the literature a framework that addresses the aforementioned optimization and interoperability challenges according to the IoT requirements. Moreover, none of the existing integrated architectures handle delay-tolerance, neither as a performance metric nor as a cost function. Our framework employs two novel ILP-based formulations that guarantee both

---

* A basic version of SIWR was introduced in Reference 13.

optimal placement for super nodes to maximize the cost-efficiency of the network and best CN selection to minimize the delay across the integrated topology while obeying link capacity and load balancing constraints. Performance of the proposed integration is evaluated and validated through extensive simulations. When compared to other existing integration schemes, our approach achieves delay drops of up to 50%. Simulation results also show that in the choice of the appropriate number of LNs/CNs, our integration can reduce the network cost by as much as 40%.

The remainder of this chapter is organized as follows. Section 5.2 surveys existing RSN integration architectures and assesses their viability in view of the challenges and constraints faced in the IoT. Section 5.2 also lists several node placement approaches in integrated networks and DTN-routing challenges. Section 5.3 presents the system models including our network model, the delay model we adopt, and our communication model. Section 5.4 describes our placement and delay-tolerant routing approaches, including our ILP formulations for super node placement and CN selection, respectively, in addition to the decentralized routing algorithm based on them. Section 5.5 presents the performance evaluation of our scheme in comparison with other integration architectures. Finally, Section 5.6 concludes this chapter.

5.2 Related Work

In this section, we investigate the main approaches to integrate WSNs and RFIDs and compare them to our integration architecture. We also assess their placement cost and efficiency with respect to IoT scenarios.

5.2.1 Architectures for Integrated RSNs

Integrated RSN architectures may follow a variety of common layouts, including (a) integrating RFID tags with sensors, (b) integrating readers with sensor nodes, and (c) mixed architectures [14,15]. We will refer to these main architectures for the remainder of the chapter as TS, RS, and MIX, respectively. All three architectures share some common components, including readers, relays, and sensor/tag nodes. The cost of sensors and tags, however, is always considered to be negligible compared to those of readers and relays that employ complicated and expensive circuitry. Thus, we quantify the cost measure of each integrated architecture by referring to the functionalities of its components as follows:

C_{Read}: Cost of reader nodes
C_{Relay}: Cost of relaying nodes
C_{XS}: Cost of extra sensor nodes in the topology that may be deployed for relaying and sensing tasks

Appending sensing capabilities to RFID tags (TS architecture) is one of the simplest ways of integration. If passive RFID tags incorporate sensors in their design, they are able to take sensor readings and transmit them to a reader as well. However, such a TS node uses the same protocols for reading tag IDs and for collecting sensed data. This option of integration limits the range of communication to RFID readers alone over single-hop links. In high-end applications, it would be extremely desirable for integrated TSs to communicate with each other as well as with other devices and form a cooperative ad hoc network (Figure 5.1).

Nonetheless, it should be noted that a system that deploys a sensor with each tagged object is highly costly and infeasible. The authors in Reference 16 propose an approach that deploys an RFID tag attached to each sensor node. It provides both unicast and multicast capability. However, this integration architecture suffers from doubling the sensing load on each integrated node. The integrated entity is required here to run at least two wireless protocols, depending on the sensed data, and perhaps some aggregation method to overcome the short communication ranges of relaying sensors that increases the system's operational and design costs, especially under large-scale deployment.

Another TS integration is to integrate sensors with active tags [17]. This option implies using batteries to power the communication circuitry of the integrated element, providing it with a longer range of communication and the ability to operate on higher rates. Nevertheless, because a battery is used, the device's cost and weight increases, while the lifetime of the RFID TSs is limited. In an IoT deployment, the incorporation of sensors into tag designs is nonrealistic and defies the advantages tags provide in terms of size and cost. Tags are usually attached to commercial goods and merchandise. It is highly improbable to include sensors, too, wherever a tag is located. Moreover, if the TS integration approach is to be delay-tolerant, then the size of the tag/sensor pairs would even increase substantially as advanced buffering options are to be included. If we define the cost of the TS architecture, it would be dominated by the total count of readers and the extra sensors used for relaying (i.e., total of $C_{Read} + C_{XS}$).

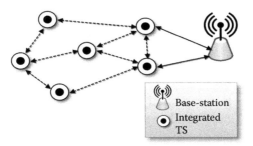

Figure 5.1   Integrated TS architecture.

The second architecture (RS) integrates RFID readers with sensor nodes. Here, the existence of three types of devices is assumed: the integrated RFID RS nodes, simple RFID tags, and the sink or base-station (Figure 5.2). Mason et al. [18] introduced a prototype system for asset tracking with RFID and sensor networks. From a DTN perspective, providing integrated entities with extra buffering capacities may not present a design challenge since readers are already complex. However, depending on integrated RFID readers and sensors to provide connectivity over the disrupted topology is not a cost-effective approach, especially when considering the limited sensing range and power consumption of sensors. As mentioned earlier, RFID readers are the most complex and costly component of an integrated system. Considering an IoT layout in which sensors are usually abundantly deployed, integrating readers with sensors will lead either to inflating the deployment costs due to the sensors' wide distribution or to depriving wide sections of the topology from the sensors' coverage for the sake of reducing the subsequent cost of integrated readers. Each of these alternatives has its toll on the system's overall performance and efficiency. Again, the cost of this architecture will be dominated by the total of $C_{Read} + C_{XS}$.

In the MIX architecture, RFID tags and sensor nodes coexist in the same network as distinct devices that are operating independently. In this architecture, depicted in Figure 5.3, the system includes three classes of devices: sensor nodes, RFID tags, and smart stations. These latter devices consist of an RFID reader, a data microprocessor, and a network interface communicating with the network's base station only. This architecture, however, is vulnerable to large-scale implementations, such as IoT settings, in which data relaying is highly required. Our proposal addresses this specific vulnerability. Also in the same regard, an architecture proposed in Reference 19 with a gateway and a sensor in a WSN

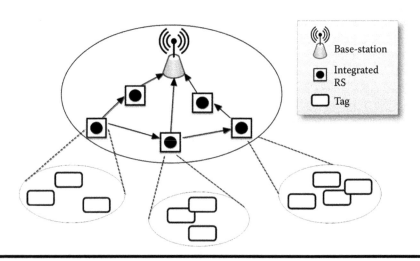

Figure 5.2   Integrated RFID RS architecture.

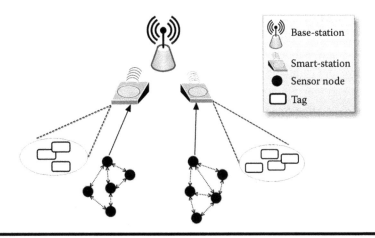

Figure 5.3   MIX architecture of RFID and WSN.

are integrated with an RFID reader and a tag, respectively. It also presents some problems related to energy imbalance among the smart nodes it introduces. Yang et al. [20] argue that in this architecture, smart nodes have a fixed transmission range. Hence, the amount of traffic that is to be forwarded increases considerably as the distance to the base station becomes shorter. Subsequently, smart nodes that are closer to the base station will run out of power earlier causing partitioning.

We remark that assuming fixed transmission ranges simply contradicts the nature of the heterogeneous devices coexisting in the IoT. In addition, we consider the architecture mentioned above to be delay-*intolerant* by design and vulnerable to partitioning, which is a concurrent nature of the IoT setting. We finally add that the proposal in Reference 19 does not specify if smart stations intercommunicate with each other, which is a desired feature to have in any integrated architecture. The authors in Reference 20 propose balancing the load among the readers by adding more readers in the area near the sink. We point out that an abundance of readers in the network may cause reader collisions, which occur when the coverage area of one RFID reader overlaps with that of another, not to mention that deploying more readers will substantially increase the cost of the system. The cost of this mixed architecture is more complicated and is express by the total $C_{Read} + C_{Relay} + C_{XS}$.

The alternative integration approach we propose, which will be further described in Section 5.4, aims toward overcoming the limitations of the architectures mentioned above in terms of delay-tolerance through the DIRSN routing scheme while controlling the cost factor through the SIWR placement scheme. This latter goal is achieved by optimally placing super nodes in a manner that guarantees the integrated RSN requirements (e.g., dissimilar coverage settings, data generation rate, and varying communication links' capacities) with the least super nodes' count and thus the least cost. This optimization deployment problem needs to be dealt with carefully. Accordingly, we discuss some of the recent placement strategies that have been proposed in the literature in this regard.

Table 5.1 summarizes the features of each of the integrated architectures mentioned above and compares them against our framework.

5.2.2 Node Placement in Integrated Architectures

Extensive work has been reported on relay node deployment strategies, which are classified into two categories: random (ad hoc) and controlled (grid-based) deployment [21–23]. These deployment approaches apply, as well, to RFID reader placement [24–27]. Nevertheless, although deploying an abundance of readers may provide full coverage, this has the side effect of creating significant interference among readers, which consequently and adversely affects the performance of the whole system. This is also not a cost-effective approach. Several maximal RFID coverage schemes have been proposed in the literature. Controlled deployment is usually pursued for indoor applications of WSNs [21]. Nevertheless, some deterministic approaches utilize algorithms to find an optimal placement of RFID readers in a grid [24], or at a predetermined set of locations [25]. While these schemes are able to achieve maximal coverage, they suffer from two major weaknesses when applied to integrated RSN layouts. First, the coverage comes at the heavy price of an abundance of integrated units. The grid approach is particularly inapplicable to integrated schemes and is artificial for real-life IoT scenarios. If the grid approach is to be applied with the RS integrated architecture, for instance, we will end up with a situation in which sensors are distributed in a grid fashion. This is a costly design that simply counters large-scale implementations of WSNs. Our proposed design, hence, avoids this by maintaining the separation of sensors and tags and focusing on the optimum distribution of reader/relay pairs (super nodes) over the network's span.

In the ad hoc placement approach, on the other hand, an algorithm needs to be executed afterwards to turn off some redundant readers. Such an approach does not eliminate the problem of readers' redundancy. Moreover, a random distribution of nodes does not solve connectivity challenges. It might even increase them. Distributing sensor nodes randomly over a given plane is a haphazard solution that does not guarantee even coverage of the sensed area. The problem grows in dimension if these nodes are mobile. Other WSN approaches such as those in Reference 28 apply sink mobility to provide an energy-efficient way for data dissemination. The proposed method navigates the mobile sink to traverse through the cluster centers according to the trajectory of an optimized route. The mobile sink then collects the data from sensors at the visited clusters.

We aim in our placement approach to specifically place readers (integrated with relays in super nodes) where they are required to perform their reading/relying double-task. Our ILP solution adopts a similar approach to the work presented in Reference 27, in which the authors propose a greedy algorithm that uses the ratio of tag counts to the number of neighboring readers of each reader to detect and eliminates

Table 5.1 A Summary of RSN Integrated Architectures Features

	Multi-Hop Communication	Transmission Load Balancing	Addresses Power Constraints	Cost-Efficient Deployment	Optimal Reader Placement	Suitable for Large-scale Deployment	Long Transmission Range	High-Rate Operation	Delay-Tolerant	Fits IoT Setting
Reference 16	✓					✓				
Reference 17	✓						✓	✓		
Reference 18	✓		✓				✓	✓	✓	
Reference 19	✓	✓						✓		
Reference 20	✓	✓	✓					✓		
SIWR & DIRSN	✓	✓	✓	✓	✓	✓	✓	✓	✓	✓

redundancy. We build on this work and develop an ILP solution that determines the optimum minimal count of reader in the layout as described in Section 5.4.

5.2.3 Data Transfer in DTNs

Data transfer, that is, routing, in DTNs has been covered extensively [29–37]. The selection of the most appropriate routing protocol depends on the level of knowledge available regarding the networks' topology. In a DTN setting, such knowledge is either partial or absolutely absent.

If the nodes possess no routing information about their neighbors, then they implement broadcasting protocols that blindly forward copies of their packets to any neighbor they encounter. This ranges from full network flooding [29,30] to partial (epidemic) flooding [31]. Network coding [32] was further proposed to improve the performance of DTN flooding. Blind routing may achieve a high delivery ratio, provided there are enough storage and energy resources. Yet, it burdens the node buffer and inefficiently utilizes the contact duration. Thus, broadcasting protocols are not favored for their costly operations. Topology control [33] is proposed to achieve efficient broadcasting with low interference and low energy consumption. Alternatively, if some knowledge of the nodes' mobility patterns or routing history is available, then guided routing [34–36] is applied. Protocols in this class exchange routing tables to assign weights to nodes/links based on information collected from the network. This information may be related to the contact between nodes or their location or mobility. For instance, history-based probabilistic routing [34,35] estimates a delivery predictability metric that is strengthened each time particular nodes meet. In addition, social-based routing [36] is particularly useful in cases in which human mobility traces are involved in the routing process.

As will be mentioned in Section 5.4, the algorithm we present here is based on exchanging routing tables between neighboring nodes. It should be mentioned, however, that routing in integrated RSNs has not been addressed in existing DTN schemes. Other DTN delivery schemes are based on utilizing the mobility patterns of nodes in the topology. We particularly note the data MULEs (DMs) approach introduced in Reference 37. DMs were defined as mobile nodes with arbitrary mobility patterns and equipped with large storage capacities and renewable energy sources. This concept is further applicable nowadays due to the wide spread of wireless smart devices. The proposal in Reference 37 consisted of a three-tier architecture (sensor nodes, DMs, and APs) and is supposed to connect sensors at the cost of high latency. This architecture resembles our network model; alas, it does not address challenges that face integrated topologies as ours does.

The previous subsections show that the existing integration approaches, as described above, have several drawbacks that cause them to be either costly and infeasible, or unsalable, thus preventing them from being efficiently deployed in real-life IoT settings. Furthermore, the architectures mentioned above assume nodal placement strategies that do not cater for cost and integrated RSN constraints, nor do they address the IoT-inherited delay-tolerance routing issues. In the remaining sections, we introduce our integrated framework that tackles these specific issues.

5.3 System Models

In this section we describe our framework's network model, the delay model for the DIRSN scheme, and the communication model used in this chapter. All three models were designed to tackle challenges of RSN applications under the IoT umbrella.

5.3.1 Network Model

We adopt a two-layered hierarchical architecture that maximizes the integrated network usability and minimizes its deployment cost. The upper layer consists of super nodes that communicate periodically with APs to deliver the data being collected at the lower layer. Super nodes are assumed to consist of RFID readers and advanced processing and communication units to aggregate sensed data and coordinate medium access, in addition to relaying data to the APs. This architecture aims toward controlling the cost factor by distributing the sensing and relaying loads over the components of integrated networks in an optimum fashion. This is a unique approach because it does not overwhelm the light sensor and tag nodes (LNs) with relaying tasks but rather focuses on the optimal distribution of the most costly reading/relaying components. Since LNs are relatively cheaper to deploy and have a minor effect on the network's cost, our scheme focuses on the upper layer's nodes (i.e., super nodes).

Figure 5.4 illustrates our RSN integration approach. This approach is based on simple assumptions related to the nature of IoT nodes:

- RFID readers have relatively low ranges of communication and are expensive. They are also not as common among the wireless devices in use in everyday settings and have to be deliberately assigned to each cluster of passive RFID tags since they are the only means of communication with these tags.
- Sensors and tags, on the contrary, are much cheaper and densely deployed over the topology but implement separate monitoring and reporting protocols.

In addition to super nodes and LNs, the IoT comprises by nature many other active nodes, including devices equipped with sensors and transceivers, which vary in their mobility, processing, communication, and buffer capacities. These resources will be exploited by our approach, as CNs, to provide connectivity whenever delay/disruption occurs.

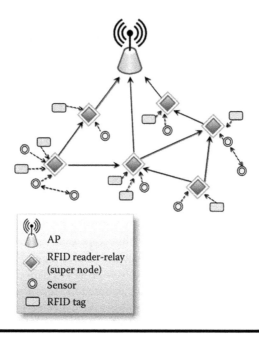

Figure 5.4   Smart integrated WSN and RFID architecture.

Figure 5.5 illustrates the basic blocks of the improved architecture upon which the DIRSN scheme is based and describes their functionalities. DIRSN assumes the existence of the following levels of nodes in a given IoT topology:

- APs that are fixed and directly connected to the Internet cloud. They act as base stations for the rest of the nodes down the hierarchy.
- Super nodes representing the integrated part of our architecture. They perform the roles of RFID readers and wireless relays to APs, simultaneously, with advanced transmission capabilities. Minimizing the number of these super nodes will minimize the deployment cost of the system as a whole since they are the most sophisticated and expensive among its components.
- LNs represented by passive RFID tags and simple sensor nodes, each performing their own protocol. LNs are assumed to be distributed densely over the topology and may be fixed or mobile depending on the application.
- CNs represented by some mobile sensor nodes (among other smart devices in the IoT) with variable buffering and transmission capabilities. A courier is potentially moving toward or residing within the communication range of an AP. These nodes buffer the data whenever connectivity is lost between super nodes and APs or between light sensors (i.e., conducting SCF routing), until a suitable "best next hop" is found. If transmission power permits, a CN may even assume the role of a relay and directly transfer a super node's load to an AP.

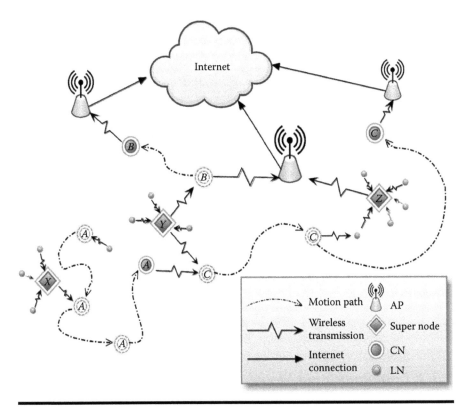

Figure 5.5 An architecture based on the DIRSN in an IoT setting.

5.3.2 Delay Model

Our objective in DIRSN is to minimize the worst delay experienced between any super node/AP pair. Thus, we adopt the delay model introduced in Reference 38. Although that model was initially proposed for underwater acoustic sensor networks, we find it generally suitable for wireless networks, regardless of the transmission medium. Most importantly, since we are using ILP, we assume its discretized delay metric that can be tuned to achieve any desired accuracy. We further propose e-health as a real-life application that justifies our architecture and its corresponding models. This will be further elaborated upon in Section 5.4.

Due to dense network topologies formulated in the IoT, a relatively long multi-hop path can easily exist between the source node and the corresponding AP. The delay components we consider in this chapter are the transmission/processing delay ψ modeled by the number of hops multiplied by ψ and the propagation delay, modeled based on the speed of signal and the Euclidian distance between two ends. The latter delay is extremely dependent on the speed of the link (or signal speed) and the

Euclidian distance between the two ends. It varies based on the utilized technology, its corresponding standards, and the transmission medium.

Accordingly, our delay model is described as follows: we define a delay step ω, which is the distance a wireless signal would travel in one time unit. Let E_{ij} be the Euclidian distance between a source node i and a destination node j; then, the discrete propagation delay over a single-hop link (i, j) would be $E_{ij}/(\omega)$. Hence, the discrete delay over a multi-hop path is the sum of the discrete delays of single-hop links that constitute that path. Note that ω (and the time unit) can be made small enough to meet any desired accuracy.

For the sake of our algorithm generality, single-hop-delay (D_{single}) and total-delay (D_{total}) can be respectively defined as follows:

$$D_{single} = \frac{E_{ij}}{\omega} + \psi \tag{5.1}$$

and

$$D_{total} = \sum_{total\ hops} \frac{E_{ij}}{\omega} + \psi \tag{5.2}$$

5.3.3 Communication Model

We assume a probabilistic model in which the probability of communication between two wireless devices decays exponentially with distance and takes into consideration surrounding obstacles and hindrances. Accordingly, the communication range of each device can be represented by an arbitrary shape. For realistic estimation of the arbitrary shape dimensions, we need a practical signal propagation model. This model can describe the path loss* in the targeted site by taking into consideration the effects of the surrounding terrain on the power (P_r) of received signals as follows [32]:

$$P_r = K_0 - 10\gamma \log(d) - \mu d, \tag{5.3}$$

where d is the Euclidian distance between the transmitter and receiver, γ is the path loss exponent calculated based on experimental data, μ is a random variable describing signal attenuation effects† in the monitored site, and K_0 is a constant calculated based on the transmitter, receiver, and field mean heights. Let P_r equal the minimal acceptable signal level to maintain connectivity. Assume γ and K_0 in Equation 5.4 are also known for the specific site to be monitored. Thus,

---

* Path loss is the difference between transmitted and received signal power.
† Wireless signals are attenuated because of shadowing and multipath effects. This refers to the fluctuation of the average received power.

a probabilistic communication model that gives the probability that two devices separated by distance d can communicate with each other is given by:

$$P_c(d,\mu) = Ke^{-\mu d^\gamma},\tag{5.4}$$

where $K_0 = 10\log(K)$.

Thus, the probabilistic connectivity P_c is not only a function of the distance separating the sensor nodes but also a function of the surrounding obstacles and terrain, which can cause shadowing and multipath effects (represented by the random variable μ).

5.4 Integrated RSN Framework

As aforementioned, our framework comprises two components: SIWR, an architecture that classifies nodes into simple LNs and integrated super nodes, and DIRSN, which is a scheme for data routing and CN selection in RSNs. The two parts aim toward providing an optimized architecture for integrated RSNs in addition to a delay-tolerant scheme for IoT settings. Particularly, we address two optimization problems related to integrate RSNs in the IoT: (1) optimal placement of super nodes and (2) locating the best set of CNs to provide minimum delay. In the following, we introduce two ILP formulations of SIWR and DIRSN, which represent our approaches to tackle the two aforementioned problems, respectively.

As previously illustrated in Figure 5.4, our scheme is based on placing LNs near the phenomenon of interest for more accurate readings. A super node is then placed on the most appropriate position to serve the largest number of LNs distributed around it based on the application's requirements. The AP is also placed with respect to the application's requirements in a fixed position and serves as the system's data sink. When compared to the integrated architectures mentioned in Section 5.2, the cost measure of our integrated architecture is expressed by the total count of super nodes, in which a single super node cost is calculated by:

$$C_{SN} = \varepsilon C_{Relay} + \eta C_{Read},\tag{5.5}$$

where ε and η denote fractional variables varying between 0 and 1 based on the hardware specifications of the designed super node. We note that Equation 5.5 combines partial costs of readers and relaying nodes which are conducted separately in the other architectures. This is due to the elimination of duplicated components achieved by our integration strategy.

The SIWR architecture, however, while providing optimum coverage to LNs, does not address challenges related to super nodes' connectivity in an IoT layout. Moreover, and most importantly, the mobility and disruptive characteristics of IoT topologies aforementioned above introduce a particular challenge regarding connecting super nodes to APs, where this connection may cease to exist for variable intervals, resulting in the isolation of entire segments of the network. Thus, we enhance this architecture in our DIRSN approach, which represents the second part of the integrated framework proposed in this chapter, by further introducing a new class of nodes (i.e., CNs) that guarantee minimum-delay connectivity between super nodes and APs (when they are out of each other's communication range), through locating the best set of CNs capable of doing so.

In the DIRSN scheme, CNs are specifically exploited to solve this connectivity challenge. In a real-life scenario, CNs would be represented by cell phones, chips on vehicles, advanced sensors, or any mobile node with sufficient transmission and buffering capabilities to exchange routing tables with neighboring nodes and discover routes to deliver data within a limited amount of time. In fact, the role of CNs in DIRSN is somehow similar to that of DMs mentioned in Section 5.2. According to our approach, however, a given CN will move across the topology according to a deterministic mobility pattern, rather than an arbitrary pattern, depending on the nature of the courier. Then, a CN will read data held by disconnected super nodes via short-range wireless communication and finally transmit the collected data to some AP in the premises. The main advantage of this approach is that it provides connectivity to otherwise disconnected super nodes. It also contributes to dedicating LNs to their sensory tasks, which consequently allows them to sustain longer life cycles. Most importantly, CNs in our architecture will obey a delay model that aims toward minimizing the maximum delay in data transfer, as we will explain in the next section.

Applications that require connectivity with constraints on delay or delivery time are common in the context of the IoT. In particular, we target emergency response systems, also known as e-health applications [39,40], which have a high demand for connectivity and a high sensitivity to failure. The use of integrated RSNs in e-health applications allows real-time monitoring of people suffering from health problems leading to earlier diagnosis. This applies as well to healthier individuals who may experience sudden health issues. Vital parameters such as heart rate, breathing rate, and blood pressure can be measured by LNs worn by the individuals to record data on multiple health parameters, allowing medical professionals to make better informed decisions. Automatic alerts can be immediately sent to medical staff to warn of deteriorations in patients' conditions or guide medical teams to the location of the case. The delay restriction of the call depends on the urgency of the situation. In the default case and since LNs are limited in their transmission capabilities, communication is initiated by neighboring super nodes or CNs on periodic bases if transmission distance permits. The readings acquired from LNs are delivered to the AP (i.e., hospital or physician's clinic) without assuming any delay constraint. However, once the readings indicate a critical condition, they are treated by the system as urgent and routes with minimum delay are computed to deliver them.

Figure 5.5 illustrates the roles of the different type of nodes through several scenarios. Note that there are three fixed super nodes: *X*, *Y*, and *Z*. There are also three mobile CNs: *A*, *B*, and *C*, in addition to several LNs, each representing an individual/patient. In this setting, the three super nodes are sufficiently placed according to SIWR to provide full coverage to any LN in the topology. As for the couriers, we see that node *A* has a mobility pattern that allows it to gather data from both super nodes *X* and *Y*. Prior to that, *A* also receives data from an LN that is assumed here to be a simple sensor, indicating that intercommunication between wireless sensors is permitted. Upon reaching the proximity of super node *Y*, which is the end of *A*'s path, node *A* realizes that it is still not in the range of any AP, so it decides to transfer its load a hop further to a neighboring courier *C*, which shows potential (that is, by exchanging routing tables with *A*) of contacting an AP in the future. *C* also reads from super node *Y*, which simultaneously forwards its urgent packets (i.e., cases to be attended by medical teams urgently) to another CN, *B*, whose routing table indicates a better probability of communicating with an AP sooner. Meanwhile, *C*'s mobility allows it to finally deliver its data load to an AP. Note that node *Z* was able to reach an AP by itself regardless of the assistance offered by any CN. Hence, it did not forward its packets to C when it was in its proximity.

Our integration framework introduces two ILP-based solutions involving optimally placing super nodes by using the SIWR approach and selecting the best set of CNs that guarantee minimum transmission delay by using the DIRSN approach. These solutions are described as follows.

5.4.1 Optimal Placement of Super Nodes

The node placement problem tackled in this chapter has an infinitely large search space, and finding the optimal solution is highly nontrivial. Therefore, we propose a model that limits the search space to a more manageable size. We assume knowledge of the terrain of the monitored site. Hence, practical candidate positions on the grid vertices can be predetermined; no feasible positions are excluded from the search space. We use cubic grid vertices to apply a novel scheme for routing and placement of super nodes in integrated WSNs and RFIDs. This scheme is used to minimize the cost of the integrated network without violating the main requirements of RFIDs and WSNs. The former requires maintaining the right ratio of tag to reader counts, while the latter requires full connectivity. Our deployment scheme, SIWR, aims at solving the following problem:

> Find the optimal locations of SN_{total} super nodes with the routing paths to deliver the generated data from each tag/sensor to the AP. The objective is to minimize the total count of super nodes.

SIWR's placement solution runs offline before the network is operational. Thus, no complexity issues are associated with it. We assume a two-layered hierarchical architecture that maximizes the integrated network usability and minimizes its deployment cost. The upper layer consists of super nodes that communicate periodi-cally with the APs (directly or via each other) to deliver the data being measured at the lower layer, composed of LNs and CNs. Super nodes are assumed to consist of RFID readers and advanced processing and communication units to aggregate received sensed data and coordinate medium access in addition to relaying mea-sured data to the APs. The SIWR architecture, which is also applied to DIRSN, aims toward dominating the cost factor by distributing the sensing and relaying loads over the components of integrated networks in an optimum fashion without overwhelming the light sensor and tag nodes (LNs) with relaying tasks but rather by focusing on the optimal distribution of the most costly integrated reading/relaying components.

The optimization problem can be formulated as an ILP. We define the following constants and variables:

Constants

- V: a set of candidate grid vertices
- v: the number of candidate positions on the grid vertices
- SN_{total}: total available super nodes
- f_{ij}: the flow from node i to node j (i.e., the data units to be sent from i to j)
- G_i: generated traffic by sensor node i
- SG_i: generated traffic by super node i
- C_i: capacity of traffic (bandwidth [BW]) available for sensor node i
- SC_i: capacity of traffic (BW) available for super node i

Variables

- α_i: a binary variable equals 1 when a sensor is placed at vertex i of the 3-D grid and 0 otherwise.
- β_i: a binary variable equals 1 when an RFID tag is placed at vertex i of the 3-D grid and 0 otherwise.
- S_i: a binary variable equals 1 when a super node is placed at vertex i of the 3-D grid and 0 otherwise.
- $N(i)$: a set of neighboring indices such that $j \in N(i)$ if node j is within the transmission range of node i (i.e., $P_c(i,j) \geq \tau$).
- $M(N(i))$: a set of indices such that $j \in M(N(i))$ if node j is within the transmis-sion range of a node that can reach one of the neighboring nodes of a node i via single or multiple hops.
- $P_c(i,j)$: the probabilistic connectivity between two nodes placed at vertices i and j.

Our policy of minimizing the cost implies minimizing the total count of SN_s without affecting the main connectivity requirements of RSN systems. By achiev-ing a placement that maintains the maximum number of RFID tags per reader, we optimally handle the main connectivity requirement in the placement of RFID systems, where a balance between energy consumption and interference is satisfied. On the other hand, we assure that each sensor will be connected to the AP through at least one path. In order to do so, we formulate the ILP as shown in Figure 5.6.

$$\text{Minimize} \quad SN_{total} \tag{5.6}$$

Subject to:

$$\sum_{i=1}^{v} S_j.\alpha_i \geq 1, \quad \forall j \in V \ \& \ j \in N(i) \tag{5.7}$$

$$\sum_{i=1}^{v} S_j.\beta_i \geq 1, \quad \forall j \in V \ \& \ j \in N(i) \tag{5.8}$$

$$S_j.\left(\sum_{j \in N(i) \& i \in (N(BS), M(N(BS)))} S_i \right) \geq 1, \quad \forall j \notin N(BS) \tag{5.9}$$

$$\sum_{i=1}^{v} S_j.\beta_i \leq OV, \quad \forall j \in V \ \& \ j \in N(i) \tag{5.10}$$

$$\sum_{j \in N(i)} \alpha_i.f_{ij} \leq C_i, \quad \forall i \in V \tag{5.11}$$

$$\sum_{j \in N(i)} S_i.f_{ij} \leq SC_i, \quad \forall i \in V \tag{5.12}$$

$$\sum_{j \in N(i)} \alpha_i.f_{ij} - \sum_{k \in N(i)} \alpha_i.f_{ki} = G_i, \quad \forall i \in V \tag{5.13}$$

$$\sum_{j \in N(i)} S_i.f_{ij} - \sum_{k \in N(i)} S_i.f_{ki} = SG_i, \quad \forall i \in V \tag{5.14}$$

$$\sum_{i=1}^{v} T_{ji} = S_j, \quad \forall j \in SNs \tag{5.15}$$

Figure 5.6   ILP formulation for the SIWR optimized placement.

Equation 5.6 is the objective function that minimizes SN_{total}. Equations 5.7 and 5.8 ensure that each sensor node is connected to at least one super node, and each tag has at least one super node in its vicinity. Equation 5.9 ensures at least one path toward the AP from a super node that is not a neighbor (i.e., having a direct communication) with the AP. Equation 5.10 guarantees that the total number of tags covered by a super node (reader) is not exceeding the optimized value (OV) of tags per reader, which has been mentioned in Reference 27. Equations 5.11 and 5.12 satisfy the traffic capacity (bandwidth) constraints. Notice that this ILP can be easily modified to handle more complex capacity constraints (by giving different weights for different links of a single node). Equations 5.13 and 5.14 guarantee the flow balance. Equations 5.11 through 5.14, together with Equation 5.10, are used to ensure the least interference in the wireless medium access. Equation 5.15 prevents the flow splitting by specifying that a super node j can transmit to only one super node i.

5.4.2 CN Selection

We apply DIRSN, a novel scheme for data routing and CNs' selection in integrated RSNs. This scheme is used to minimize the delay of the integrated IoT network without violating the main requirements of RFIDs and WSNs. DIRSN aims at solving the following problem:

> Given a large search space (modeled by IoT random deployment) and a limited number of super nodes and static/mobile LNs and CNs, find the optimal set of CNs with their corresponding routing paths to deliver the generated data from each super node covering tag/sensor LNs to the AP in minimum delay based on current network topology.

We assume discretized rounds by which our scheme selects routing paths based on locating the most suitable couriers passing through these paths. This is determined according to minimum delay and link constraints calculated from the lower layer of the architecture (LNs) up to super nodes.

Figure 5.7 shows an example in which four super nodes (A, B, C, and D, respectively) have established at some point in time one or more connecting paths through some CNs to one AP. Note that each super node–AP path has its end-to-end delay and capacity characteristics as speculated by the SIWR ILP given in the format (delay/capacity). These characteristics are based on table exchanges between CNs and super nodes. The ILP of DIRSN selects the path (illustrated in the thick red line) that guarantees minimum delay without violating capacity constraints. The decision of utilizing this CN depends on its resources, which are stored at the super node's routing table. For example, assuming that super node A in Figure 5.7 has to transmit packets with a load-balance constraint that requires a minimum link capacity of 10 MB, the ILP detects three different paths connecting A to the

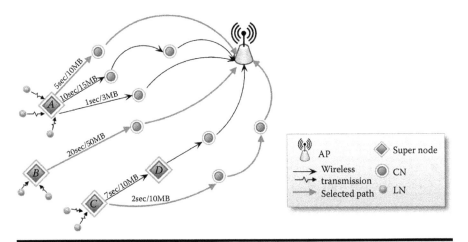

Figure 5.7   Example of DIRSN path selection.

AP. Two of the three paths obey the capacity constraints, and the one among these two that provides the minimum delay is ultimately chosen. Note that the third path does provide minimal delay from A to the AP. However, it fails to fulfill the required capacity; hence, it was not chosen by the ILP. The case is easier with super node B, for which there is only a single path connecting it to the AP. The ILP is forced to select this path, regardless of its promised delay once it fulfills the capacity constraint. As for super node C, there are two paths with equal link capacities linking it to the AP, one of which passes through another super node, D. Obviously, the path that provides minimum delay is selected by the ILP in this case.

The selection of the best M couriers in the proximity of each super node and available CN in the network is achieved by periodically exchanging routing tables and/or registration records between neighboring nodes. This selection process is repeated at the beginning of each triggered round and will be explained later. The mobility history of the neighbors is examined against the communication range of the corresponding destination node(s). Based on the results, forwarding candidates are defined according to best (i.e., minimum) delays. The algorithm we use to conduct this approach is based on an ILP formulation that, in turn, requires defining the following constants and variables:

Constants

- C_{total}: total available couriers for data transfer
- D_{max}: maximum delay to transfer a data unit from a super node to the courier
- SG_i: a data generation rate of a super node i (based on the underlining connected LNs per super node)

- t_i: the traffic capacity of a super node i (i.e., maximum data units that can be relayed by a super node per round)
- T_i: the traffic capacity of a CN i (i.e., maximum data units that can be relayed by a courier per round)
- E_{ij}: the Euclidian distance between node i and j
- w: the distance a node signal would travel in one time unit
- M: the candidate couriers' count available per round
- $N(i)$: a set of neighboring candidates such that $j \in N(i)$ if node j is within the transmission range of node i (i.e., $P_c(i,j) \geq \tau$)
- $M(i)$: a set of indices such that $j \in M(i)$ if node j is within the range of a CN i that can reach an AP

Variables

- c_i: a binary variable equal to 1 when a CN at position i (associated with an (x,y) coordinate) is chosen by a super node to relay its data to the AP and equal to 0 otherwise
- f_{ij}: the flow from super node i to CN j (i.e., the data units to be sent from i to j)
- l_{ij}: the flow from a CN i to CN j

Our policy of minimizing the delivery delay implies minimizing the total path length toward the AP without overwhelming the integrated network. This is achieved by locating a courier set that maintains the shortest path from each super node to an AP while considering their varying node/link capacities and load balance. We aim for each super node to deliver data to its corresponding AP with the least delay. In order to do so, we formulate the ILP in Figure 5.8. Equation 5.16 is the objective function that minimizes D_{max}. Equations 5.17 and 5.18 satisfy the traffic capacity constraints available to super nodes and CNs, respectively. Equation 5.19 guarantees the flow balance and controls data generation rates at super nodes. Equation 5.20 guarantees that if no courier is selected (i.e., $C_j = 0$), no flow is sent to courier at position j. Equation 5.21 makes D_{max} the maximum delay over all super nodes seeking the AP (note that we minimize D_{max}). Equation 5.22 satisfies the constraint that only M couriers are available. We remark that this ILP can be modified to handle more complex capacity constraints (e.g., given different weights to different links incident to a single super node and/or CN). We show a general case here for ease of the presentation.

We define Algorithm 5.1 to find candidate couriers that guarantee minimum delay transmission to APs among CNs and super nodes and to execute the ILP formulated in DIRSN. A link is only chosen if it offers minimum delay and satisfies the capacity constraint that is dependent on the packet generation rate per LN. End-to-end paths are determined based on the updates of routing tables that take place between super nodes and CNs.

$$\text{Minimize} \quad D_{max} \tag{5.16}$$

Subject to:

$$\sum_{j \in N(i)} f_{ij} \leq t_i, \qquad 1 \leq i \leq SN_{total} \tag{5.17}$$

$$\sum_{j \in N(i)} f_{ij} + \sum_{j \in M(i)} l_{ij} \leq T_i, \qquad 1 \leq i \leq C_{total} \tag{5.18}$$

$$\sum_{j \in N(i)} f_{ij} - \sum_{j \in N(i)} f_{ji} = SG_i, \qquad 1 \leq i \leq SN_{total} \tag{5.19}$$

$$\sum_{i \in M(j)} l_{ij} \leq c_j \sum_{1 \leq i \leq SN_{total}} SG_i, \qquad 1 \leq j \leq C_{total} \tag{5.20}$$

$$\sum_{j \in N(i)} D_{single} \cdot f_{ij} + \sum_{j \in M(k)} D_{single} \cdot l_{kj} \leq D_{max}, \quad 1 \leq i \leq SN_{total}, 1 \leq k \leq C_{total} \tag{5.21}$$

$$\sum_{i=1}^{C_{total}} c_i = M \tag{5.22}$$

Figure 5.8   ILP formulation for the DIRSN.

Algorithm 5.1: (DIRSN approach): Finding candidates for minimum delay to APs among couriers and super nodes

Procedure FindMinDelayCandidates()
set $Delay_{max} = \infty$
do for each triggered round
    for each *super node* SN_i and *courier node* CN_i **do**
    find all neighbours
        if $v_{neighbour} \leq V_x$
        exchange routing tables
    for each *access point* AP_i
        find set of neighbours with minimum $Delay_{APi} < Delay_{max}$
            set $Delay_{max} = Min (Delay_{APi}$
    end

    end
end
if (new data to forward && $v_{neighbour} \leq V_x$)
      call DIRSN ILP
end

Note that the DIRSN algorithm is applied only subject to the following conditions: (a) the availability of new data to be forwarded and (b) the availability of suitable neighboring CNs. A suitable courier is one that passes by the super node at a speed that does not exceed a threshold V_x, which is equivalent to pedestrians and low-speed mobile objects. Also note that until the results of the ILP are returned, the DIRSN algorithm will use existing routes. While this may result in utilizing suboptimal routes, our simulation results (as discussed in Section 5.2) show that the algorithm is still superior to existing solutions. It should be also noted that the ILP in the DIRSN algorithm is executed in super nodes and CNs, which are supposedly more computationally capable nodes.

5.5 Discussion and Results

In this section, we evaluate the performance of our proposed framework in practical settings, where the aforementioned probabilistic communication model is considered. We compare our proposed SIWR and DIRSN approaches, respectively, against the three RSN integration architectures discussed earlier in Section 5.2 (namely TS, RS, and MIX). These architectures represent our comparison baseline in this study. In fact, simplified variations of these architectures were widely studied in the literature [14–20].

To compare the performance of the integrated RSN architectures, the following four performance metrics are used: cost, average delay, average packet loss, and average generation rate (AGR). The first metric, cost, is used in the SIWR simulations. Cost represents the total count of relays, readers, and/or super nodes in a given architecture multiplied by their cost measure previously mentioned in Section 5.2 for each architecture type. We remark that the super nodes count in case of TS, RS, and MIX architectures will be the count of total readers and sensors relaying data to the APs.

The other three performance metrics are used in the DIRSN simulations. Unless otherwise specified, we use the average delay as a measure of latency. Delay is measured in msec and is defined as the amount of time required to deliver a data unit to the AP. The average packet loss percentage, on the other hand, is the percentage of transmitted data packets that fail to reach the AP reflecting the effects of bad communication channels on data delivery over the architecture. Finally, AGR is measured in Kbps. It is represented by the traffic rate generated by each super node or any equivalent node in a given architecture. In fact, AGR can be treated as a representation of the traffic generated by LNs in the network since it is a reflection of LNs count. In case of TS, RS, and MIX architectures, the AGR at super nodes will be the AGR of readers and sensors relaying data to the couriers.

While studying these performance metrics, we vary three main parameters: LN count (in SIWR simulations), CN count, and network cost (in DIRSN simulations). The LN count reflects the scalability of the utilized architecture under varying application scales. In addition, it reflects the application's complexity. The CN count reflects the scalability of the exploited architecture under varying application scales. In addition, it reflects the application's complexity.

5.5.1 Simulation Model

The proposed mixed integer linear program (MILP)-based approach is applied to the different integrated RSNs architectures by using MATLAB. Based on experimental measurements [32], we set the communication model variables to be as follows: $\gamma = 4.8$, $Pr = -104$ (dB), $K_0 = 42.152$, and μ to be a random variable that follows a log-normal distribution function with a mean of 3 and variance of 10. We choose $\varepsilon = \eta = 0.8$ according to Reference 41 in order to consider the most expensive super nodes. All mobile nodes are set to follow the random waypoint mobility model [42]. We set the V_x value to be 5 km/h, which allows couriers represented by pedestrians to be considered in e-health applications. The MATLAB simulator determines whether a wireless node is connected to its neighbors or not based on the aforementioned probabilistic communication model, where $\tau = 70\%$. Each simulation experiment is repeated 500 times, and the average results hold a confidence interval no more than 5% of the average (over 500 runs) at a 95% confidence level.

5.5.2 Simulation Results

The results as plotted in Figures 5.9 through 5.11 show that our SIWR architecture indeed outperforms the other three RSN integration architectures and always requires fewer super nodes. This translates into minimal system cost and proves the optimality of our deployment solution proposed by SIWR.

In our first ILP iteration, we assumed a density of three LNs per m². As shown in Figure 5.9, the total cost increases linearly according to the four architectures as the count of LNs does, as to be naturally expected. This interesting observation is related to our SIWR architecture, which maintains the least cost throughout the iteration, leading to significant cost savings (more than 35% in comparison to the best of the other architectures). The same observation, with even better performance (i.e., much lower cost required by SIWR), stands when increasing LNs' density to 30 nodes per m², as shown in Figure 5.10. However, when this density is increased to 300 LNs per m² and the total LNs' count in the topology reaches the range of thousands, which is trivial in many large-scale applications involving WSNs and RFIDs, the performance differences between SIWR and the competing integration architectures become astonishingly obvious. As shown in Figure 5.11, when the

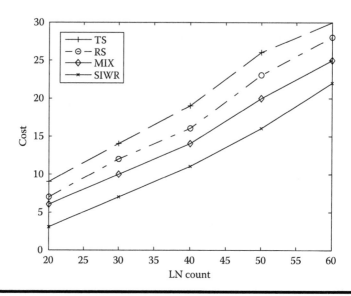

Figure 5.9 Cost comparison with three LNs/m².

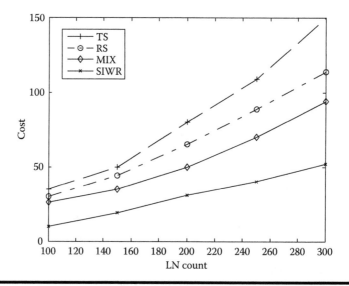

Figure 5.10 Cost comparison with 30 LNs/m².

count of LNs exceeds 2000, the corresponding network cost increases exponentially, as does the deployment cost of the involved architectures. This is not the case with SIWR, which maintains a much steadier rate of cost increase that remains definitely lower than any of the other architectures. We attribute this steadiness to

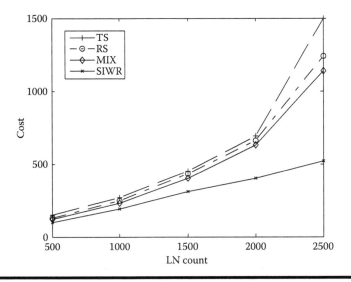

Figure 5.11 Cost comparison with 300 LNs/m².

the SIWR's ILP that guarantees a minimal deployment of super nodes and thus dominates the cost factor as opposed to the other architectures that tend to increase the number of relays or readers redundantly as the number of LNs increase in their topologies in order to assure connectivity at a very costly deployment price.

As for the DIRSN simulations, we apply its ILP-based approach while increasing the couriers' count from 10 to 70. This range of couriers is applied to DIRSN and the other three integration architectures simultaneously, and the resulting average delay in each of the four architectures is plotted against the corresponding count. To better understand maximum and minimum delays, we provide statistics of median, minimum, and maximum experienced delays (see Figure 5.12). In Figure 5.12, top and bottom ends of boxes represent the 75th and 25th percentiles of delay, respectively. Top and bottom extensions of the boxes represent the 95th and 5th percentiles of delay, respectively. We conclude from Figure 5.12 that delay values are very densely distributed condensed. Thus, that average delay is sufficient to give a clear understanding of this particular metric.

Consequently, Figure 5.13 shows that for all integrated architectures, the average delay tends to drop rapidly as the couriers' count increases. This is only natural since an abundance of couriers will surely increase the connectivity over any network. We note, however, that even with as few as 10 couriers in the layout, our DIRSN approach scores an average delay that is 30% better than the best average delay achieved by the other architectures. As the number of couriers increases to 70, an average delay drop of up to 50% is achieved by DIRSN. We attribute this to DIRSN's distinct delay-tolerant courier selection algorithm.

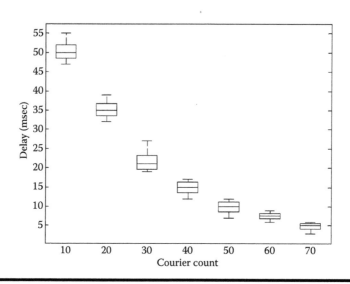

Figure 5.12   Delay statistics per number of couriers (DIRSN).

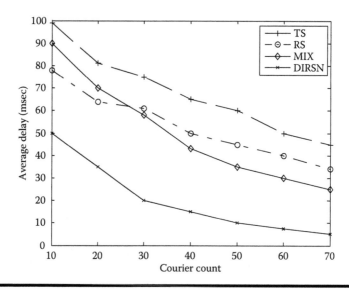

Figure 5.13   Average delay per number of couriers.

In Figure 5.14, we apply the same method while increasing the count of super nodes from 10 to 70 and applying this to all four architectures. Super nodes in DIRSN are responsible for RFID coverage in addition to performing relay transmissions. An increase of the super node count apparently lowers the average delay for all four architectures, but this count increase, as we mentioned earlier, adds to the cost of the network. Figure 5.14 confirms the cost efficiency of our integration

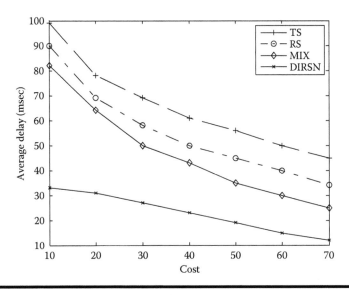

Figure 5.14   Average delay per number of super nodes.

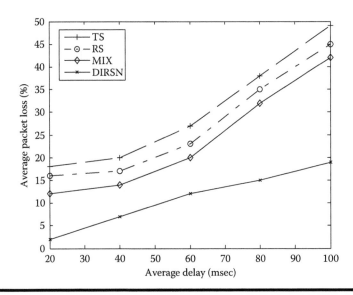

Figure 5.15   Average packet loss against average delay.

approach. It indicates that the best among the other three rival architectures (i.e., MIX) has to deploy 70 super nodes to achieve an average delay that the DIRSN can accomplish with only 10 super nodes.

Figure 5.15 compares average delay against our third performance metric, that is, average packet loss percentage. Again, DIRSN produces the best (minimal)

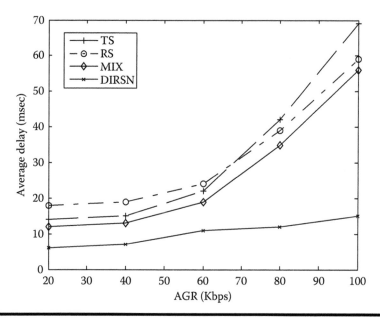

Figure 5.16 Average delay compared AGR.

packet loss as delay increases (almost 60% of the packets are saved). This is trans-lated to better coverage and connectivity in our architecture compared to others. In other words, DIRSN is more successful in providing alternative paths to deliver messages to their final destinations than the rest of the integrated architectures. This reflects the optimality of our placement and selection ILPs. Average delay and average packet loss are always directly proportional. Yet, according to Figure 5.15, DIRSN provides a steadier and linear rate of delay increase that remains substan-tially lower than the delay increase of the other three architectures, which all show an exponential rise in average packet loss beyond the 60-msec delay mark.

Finally, we compare the average delay to the AGR measured in Kbps. AGR is the traffic rate generated by the LNs that are associated to super nodes in the net-work. As the delay increases, AGR does as well. However, Figure 5.16 also shows that DIRSN produces an increase rate that is superbly steadier (linear) than the other architectures that experience, again, an exponential increase in their average delays as their corresponding AGR increases.

The results as plotted in Figures 5.13 through 5.16 show that our DIRSN archi-tecture indeed outperforms all the three common RSN integration architectures. It is worth mentioning, however, that through the displayed results, MIX archi-tecture always comes as a second best to DIRSN, with a considerable gap between the two. We explain this as a result of MIX being the closest integration architec-ture to DIRSN in terms of treating sensors and tags as separate entities. DIRSN, however, excels in adopting an integration approach that is cost-efficient in terms of optimally deploying super nodes and locating CNs in a way that serves delay, delivery, and cost constraints better and, as shown above, produces results that surpass other architectures.

5.6 Conclusion

In this chapter, we introduce an optimized delay-tolerant framework for integrated RSNs in IoT. Our framework is made up of two components. The first is SIWR, a novel, smart, integrated WSN and RFID architecture that classifies nodes into LNs and integrated super nodes. The integration choice is meant to relieve LNs, that is, simple sensors and tags, from any relaying load and, hence, concentrating the cost factor of our architecture in the optimal deployment of the super nodes. This particular task was catered for by an ILP solution that minimizes the overall cost of the system by minimizing the total count of super nodes without affecting the main connectivity requirements of RSN systems.

The second part of our framework tackles the intermittent connectivity challenge facing integrated IoT layouts and is represented by DIRSN, an optimized delay-tolerant approach for integrated RSNs. This is a novel scheme for data routing and couriers' selection in RSNs. Within our framework, DIRSN further enhances SIWR's architecture to better suit node characteristics in the IoT paradigm. The DIRSN scheme introduces a class of mobile CNs that perform delay-tolerant probabilistic SCF routing and are responsible for providing minimum-delay connectivity between super nodes and their corresponding APs. DIRSN's ILP solution is triggered by the need to deliver data, given the availability of suitable couriers and subject to capacity measures. It minimizes delay across the network without violating the main RFID- and WSN-dense deployment and capacity requirements, such as load balancing, and builds on SIWR's novel architecture to locate the best set of CNs that promise to provide connectivity. Hence, the ILPs of our framework are intended to provide RFID coverage and full WSN connectivity, respectively, for integrated RSNs in IoT settings.

We compared our integrated framework with three other integration models for RSNs, namely TS, RS, and MIX [14,20]. The comparison was based on four performance metrics cost, average delay, average packet loss, and AGR. Throughout all the simulation runs, our integration approach substantially outperforms the other architectures. Our integration framework ensures better delivery rates and lesser delay intervals than any existing integration approach, with a cost efficiency that is not preceded by them. This is confirmed by our simulation results that show that SIWR and DIRSN require the least number of super nodes to achieve delivery and delay results better than the other architectures. We conclude that our framework and its two ILP formulations represent the optimal delay-tolerant transfer architecture for RSNs in IoT in terms of delay and connectivity constraints.

Acknowledgments

This research is partially funded by a grant from the Ontario Ministry of Economic Development and Innovation under the Ontario Research Fund-Research Excellence (ORF-RE) program. This research is also supported by the National Plan for Science and Technology at King Saud University, project number: 11-INF1500-02.

References

1. INFSO D.4 Networked Enterprise & RFID INFSO G.2 Micro & Nanosystems, in Cooperation with the Working Group RFID of the ETP EPOSS, *Internet of Things in 2020: A Roadmap for the Future, Version 1.1*, Brussels, Belgium: European Commission, May 27, 2008.
2. H. Sundmaeker, P. Guillemin, P. Friess and S. Woelffle, *Vision and Challenges for Realising the Internet of Things, Cluster of European Research Projects on the Internet of Things*, Brussels, Belgium: European Commision, 2010.
3. D. Yang, F. Liu, and Y. Liang, *A Survey of the Internet of Things, in ICEBI-10, Advances in Intillegant Systems Research*, ISBN 978-90-78677-40-6, Atlantis Press, December 2010.
4. L. Atzori, A. Iera, and G. Morabito, The internet of things: A survey, *The International Journal of Computer and Computer Networks*, vol. 54, no. 15, pp. 2787–2805, 2010.
5. G. Santucci, The Internet of things: Between the revolution of the Internet and the metamorphosis of objects, H. Sundmaeker, P. Guillemin, P. Friess, and S. Woelffle, Eds., *Forum American Bar Association*, pp. 11–24, 2010, http://cordis.europa.eu/fp7/ict/enet/publications_en.html
6. HP Labs, Central Nervous System for the Earth (CeNSE), 2010, http://www.hpl.hp.com/research/intelligent_infrastructure
7. J. Bleecker. A Manifesto for Networked Objects—Cohabiting with Pigeons, Arphids, and Aibos in the Internet of Things, blog, 2006. http://www.nearfuturelaboratory.com/files/WhyThingsMatter.pdf
8. E. Welbourne et al. Building the internet of things using RFID: The RFID ecosystem experience, *IEEE Internet Computing*, vol. 13, no. 13, pp. 48–55, 2009.
9. G. Kortuem, F. Kawsar, D. Fitton, and V. Sundramoorthy, Smart objects as building blocks for the internet of things, *IEEE Internet Computing*, vol. 14, no. 1, pp. 44–51, 2010.
10. H. Nakayama, Z. Fadlullah, N. Ansari, and N. Kato, A novel scheme for WSAN sink mobility based on clustering and set packing techniques, *IEEE Transactions on Automatic Control*, vol. 56, no. 10, pp. 2381–2389, 2011.
11. K. Fall, A delay-tolerant network architecture for challenged internets, in *ACM SIGCOMM'03, Proceedings of the 2003 Conference on Applications, Technologies, Architectures, and Protocols for Computer Communications*, Karlsruhe, Germany, pp. 27–34, 2003.
12. J. Wu, S. Yang, and F. Dai, Logarithmic store-carry-forward routing in mobile ad hoc networks, *IEEE Transactions on Parallel and Distributed Systems*, vol. 18, no. 6, pp. 735–748, 2007.

13. F. Al-Turjman, A. Al-Fagih, and H. Hassanein, A novel cost-effective architecture and deployment strategy for integrated RFID and WSN systems, in *Proceedings of IEEE International Conference on Computing, Networking and Communications,* pp. 835–839, 2012.

14. H. Liu, M. Bolic, A. Nayak, and I. Stojmenovie, Integration of RFID and wireless sensor networks, in *Proceedings of Sense ID 2007 Worksop at ACN SenSys,* Sydney, Australia, 2007.

15. A. Mitrokotsa and C. Douligeris, *Integrated RFID and Sensor Networks: Architectures and Applications,* ISBN 978-14-20077-77-3, Auerbach Publications, CRC Press, Taylor & Francis Group, New York, USA, Chapter 18, pp. 511–535, 2009.

16. A.G. Ruzzelli, R. Jurdak, and G.M.P. O'Hare, On the RFID Wake-up impulse for multi-hop sensor networks, in *Proceedings of 1st ACM Workshop on Convergence of RFID and Wireless Sensor Networks and their Applications (SenseID),* Sydney, Australia, 2007.

17. A. Ferrer-Vidal, A. Rida, S. Basat, L. Yang, and M.M. Tenzeris, Integration of sensors and RFID's on ultra-low-cost paper-based substrates for wireless sensor networks applications, in *Proceedings of IEEE WiMesh,* Reston, VA, USA, pp. 126–128, 2006.

18. A. Mason, A. Shaw, and A.I. Al-Shamma'a, Asset tracking: Beyond RFID, in *7th Annual PG Symposium on the Convergance of Telecommunications,* Networking and Broadcasting, Liverpool, UK, pp. 267–272, 2006.

19. L. Zhang and Z. Wang, Integration of RFID into wireless sensor networks: Architectures, opportunities and challenging problems, in *Proceedings of GCCW,* Hunan, China, pp. 463–469, 2006.

20. G. Yang, M. Xiao, and C. Chen, A simple energy-balancing method in RFID sensor networks, in *Proceedings of IEEE International Workshop on Anti-Counterfeiting, Security, Identification,* Xiamen, China, pp. 306–310, 2007.

21. F. Al-Turjman, H. Hassanein, W. Alsalih, and M. Ibnkahla, Optimized relay placement for wireless sensor networks federation in environmental applications, *Wiley: Wireless Communication & Mobile Computing Journal,* vol. 11, no. 12, pp. 1677–1688, 2011.

22. M. Younis and K. Akkaya, Strategies and techniques for node placement in wireless sensor networks: A survey, *Ad Hoc Networks,* vol. 6, no. 4, pp. 621–655, 2008.

23. D. Yang, S. Misra, and G. Xue, Joint base station placement and fault-tolerant routing in wireless sensor networks, *IEEE GLOBCOM,* Honolulu, HI, USA, 2009.

24. A. Oztekin, F. Mahdavi, K. Erande, Z. Kong, L.K. Swim, and S.T.S. Bukkapatnam, Criticality index analysis based optimal RFID reader placement models for asset aracking, *International Journal of Production Research,* vol. 48, no. 9, pp. 2679–2698, 2010.

25. L. Wang, B.A. Norman, and J. Ragopal, Placement of multiple RFID reader antennas to maximize portal read accuracy, in *International Journal of Radio Frequency Identification Technology and Applications,* vol. 1, no. 3, pp. 260–277, 2007.

26. A. Al-Fagih, F. Al-Turjman, H. Hassanein, and W. Alsalih, Coverage-based placement in RFID networks: An overview, *3rd FTRA Intl. Conference on Mobile, Ubiquitous and Intelligent Computing (MUSIC-12),* Vancouver, Canada, pp. 220–224, 2012 (Accepted).

27. K. Ali, W. Alsalih, and H. Hassanein, Using neighbor and tag estimations for redundant reader eliminations in RFID networks, in *IEEE Wireless Communications and Networking Conference (WCNC),* Cancun, Mexico, pp. 832–837, 2011.

28. H. Nakayama, N. Ansari, A. Jamalipour, and N. Kato, Fault-resilient sensing in wireless sensor networks, *Computer Communications*, vol. 30, no. 11–12, pp. 2375–2384, Sept. 2007.

29. K.A. Harras, K.C. Almeroth, and E.M. Belding-Royer, Delay tolerant mobile networks (DTMNS): Controlled flooding in sparse mobile networks, *Lecture Notes in Computer Science*, vol. 3462, pp. 1180–1192, 2005.

30. T. Spyropoulos et al., Spray and wait: An efficient routing scheme for intermittently connected mobile networks, in *Proceedings of the 2005 ACM SIGCOMM Workshop on Delay-tolerant Networking*, Philadelphia, Pennsylvania, USA, pp. 252–259, 2005.

31. A. Vahdat and D. Becker, *Epidemic Routing for Partially Connected Ad Hoc Networks*, Technical Report CS-200006, Department of Computer Science, Duke University, Durham, NC, 2000.

32. Y. Wang, S. Jain, M. Martonosi, and K. Fall, Erasure-coding based routing for opportunistic networks, in *Proceedings of the 2005 ACM SIGCOMM Workshop on Delay-tolerant Networking*, Philadelphia, Pennsylvania, USA, pp. 229–236, 2005.

33. K. Miyao, H. Nakayama, N. Ansari, and Nei Kato, LTRT: An efficient and reliable topology control algorithm for Ad-Hoc networks. *IEEE Transactions on Wireless Communications*, vol. 8, no. 12, pp. 6050–6058, 2009.

34. A. Lindgren, A. Doria, and O. Schelen, Probabilistic routing in intermittently connected networks, in *Lecture Notes in Computer Science*, 2004, pp. 239–254.

35. J. Burgess, B. Gallagher, D. Jensen, and B.N. Levine, MaxProp: Routing for vehicle-based disruption-tolerant networks, in *INFOCOM 2006. 25th Proceedings of the IEEE International Conference on Computer Communications*. Barcelona, Spain, 2006.

36. P. Hui, J. Crowcroft, and E. Yoneki, Bubble-Rap: Social-based forwarding in delay tolerant networks, in *IEEE Transactions on Mobile Computing*, vol. 10, no. 11, 2011, pp. 1576–1589.

37. R. Shah, S. Roy, S. Jain, and W. Brunette, Data MULEs: modeling a three-tier architecture for sparse sensor networks, in *IEEE SNPA Workshop*, Anchorage, AK, USA, 2003.

38. W. Alsalih, S. Akl, and H. Hassanein, Placement of multiple mobile data collectors in underwater acoustic sensor networks, in *Proceedings of IEEE International Conference on Communications (ICC)*, Beiging, China, May 2008.

39. J. Rodrigues, S. Fraiha, H. Gomes, G. Cavalcante, A. de Freitas, and G. de Carvalho, Channel propagation model for mobile network project in densely arboreous environments, *Journal of Microwaves and Optoelectronics*, vol. 6, no. 1, pp. 236–248, 2007.

40. C. Swedberg, Dutch researchers focus on RFID-Based sensors for monitoring Apnea, Epilepsy, *RFID Journal*, December 2007, http://www.rfidjournal.com/article/articleview/3780/1/1

41. D. Ranasinghe, M. Sheng, and S. Zeadally, *Unique Radio Innovation for the 21st Century: Building Scalable and Global RFID Networks*, First edition, ISBN 978-3-642-03461-9, Springer-Verlag, UK, 2011.

42. T. Camp, J. Boleng, and V. Davies, A survey of mobility models for ad hoc network research. *Wireless Communication & Mobile Computing (WCMC): Special issue on mobile Ad Hoc networking: Research, Trends and Applications*, vol. 2, no. 5, 2001, pp. 483–502.

Chapter 6

Multimedia-Enabled WSNs Using UAVs for Safety-Oriented Mobile IoT

6.1 Introduction

Drones, also known as unmanned aerial vehicles (UAVs), have been used mainly in military applications for many years. However, there has been a recent increase in the use of UAVs in nonmilitary fields, which is inspired by the 5G revolution. Such fields include precision agriculture, security and surveillance, delivery of goods, and provisioned services [1]. For example, Amazon and Walmart have been working on a new system to deliver goods to customers over them air. Additionally, China's largest mailing company, DHL, has started delivering around 500 parcels daily using UAVs. Moreover, we can use some 5G-supported UAVs to monitor and send feedback from incidents that happen along the road, hence eliminating road support teams. Moreover, a traffic policeman can be replaced or assisted by a UAV by hovering over fast-moving vehicles and reporting back traffic violations. Consequently, the use of UAVs for industry-oriented services may become a reality very soon, especially after the revolution of communication systems toward realizing the 5G-inspired Internet of things (5G/IoT) paradigm. A key field of interest for the IoT and sensor networks is the development of wearables that can connect to these UAVs for various application areas. Having an infrastructure such as 5G that is developed with consideration of IoT applications in details causes a significant need for further contributions in terms of data and especially multimedia delivery. 5G is targeting 10-Gbps data rates in real-time networks. The recent tests at the 5G Innovation Centre have shown that it is even possible to exceed 1 Tbps in a laboratory environment. This would mean being able to transmit 33 high-definition films, each up to 2.5–3 hours in running length, in a single second. It is typically desirable to have these infra-structures seamlessly integrated with the IoT industrial solutions, and there are recommended prototype sensors for similar applications with the energy and marginal cost for each added sensor.

Wireless sensor networks (WSNs) are very critical in the aforementioned archetype. The integrated 5G and IoT is termed as an extraordinarily complex model in which devices are deployed as consumer elements forming a complex interconnected system. Conversely, these elements operate with very strict energy constraints, resulting in the amount energy left over for fault-tolerance procedures being very limited. Moreover, the emergence of the variety of multimedia IoT appli-cations, such as video streaming from smart homes, smart cities, public intelligent transportation systems, health care monitoring via public cameras, etc., will most certainly increase the need for a fault-tolerant routing procedure in different ways, because of the desired sustainable mode of operation, for a successful multimedia delivery process [2].

Nowadays, WSNs function in an autonomous manner with a very lim-ited human control in a UAV-enabled system, in which sensors and cameras are attached/distributed not only in smart environments but also to flying UAVs in the industry. Moreover, most of these sensors are positioned in wild outdoor envi-ronments and sometimes even harsh environments. Hence, it is quite difficult to determine and design a fault-tolerant routing protocol. Because the communica-tion energy is considerably lower than that used in computations, it is very impor-tant to come up with a fault-tolerant algorithm that is able to recover from path failures, no matter what the added computational energy is. If this is not done, any random event may cause the UAV to fail in delivering its exchanged information and interrupt network functionality.

Accordingly, this necessitates a multipath routing approach that can recover the failed path. Multipath routing protocols form a good candidate for a more reli-able 5G/IoT paradigm, in which fault-tolerance routing problems are considered as optimization problems. These optimization problems formulate a *k*-disjoined path to encounter up to *k*-1 path failure. Exceptional fault-tolerance routing in UAV-enabled networks needs humongous computational power, which brings about large control message overhead without scalability as the problem increases [3]. Coming up with a solution to these problems on each sensor may require signifi-cant capacities in terms of memory and computational resources, and it still may only produce ordinary results.

To offer quick recovery from failures, we have designed a bio-inspired routing algorithm called particle swarm optimization (PSO). The authors in Reference 4 note that the use of PSO has produced positive results due to its simple concept and high efficiency. Nonetheless, despite competitive performance, there still remains a huge challenge of solving the fault-routing problem because of the convergence issue. However, many of the impulsive convergence traps occur due to fast convergence features and a diverse loss of particle smarm, and hence, it results in adverse solutions. In addition, the ability to differentiate between exploration and exploitation search is another significant challenge that we face today. Exploration contains the swarm convergence, whereas exploitation usually tends to make the swarm particle convergence without leaving the viable area that eventually leads to premature convergence; hence, it is never proper to overemphasize exploitation or exploration [5]. Due to these challenges, and especially connectivity issues, we propose a new approach that is more efficient in recovering failures via multipath routing capable of attaining quality of service (QoS) in terms of energy consumption, lifetime, delay, and throughput. The proposed multipath routing algorithm is compared against existing optimization algorithms, namely the canonical particle swarm (CPSO) [6], fully multi-swarm (FMPSO) [7], and multi-swarm particle swarm optimization (CMPSO) algorithms [5] to offer a different learning technique for the swarm particles. The aforementioned algorithms are only different from each other in that they have different learning contrivances and the like; otherwise, they are similar to each other. Additionally, increasing the number of paths requires more message exchange and communication overhead [6]. Therefore, we adopt the use of intricate network connections so as to denote layout of the swarm and use the multipath routing algorithm to stabilize the trade-off between fault tolerance and communication overheard by taking advantage of a mixed proactive and reactive routing mechanism that maintains the best objective function value for the designated paths per particle. Afterwards, the particles are increased or decreased and then, given a velocity that suits them, the augmented objective function must be used in order to make a fitting assortment.

Due to the aforementioned issues in WSN technology, and especially the connectivity ones, we propose a new routing algorithm that is more efficient in considering multipath failures that contain reconstructive procedures capable of attaining QoS in terms of network lifetime, energy consumption, delay, and throughput.

6.2 Related Work

Apart from the fact that fault-tolerant routing makes the network system more reliable, it is also very important when it comes to 5G or IoT since WSNs heavily depend on surrounding environments to interact with; hence, the need to provide QoS is a must. There are different ways to determine a fault-tolerant route in WSNs; one of the leading ways is multipath routing [8]. The authors in Reference 9 claim that recent optimization methods (including meta-heuristic ones) are more effective in multipath routing. Moreover, the author in Reference 10 comes up with a solution to the disjoined multipath problem by proposing a new energy-efficient multipath routing system based on using PSO. The author in Reference 11 introduces PSO that is used to select routes for load delivery. On the other hand, the author in Reference 12 recommends an enhanced PSO-based clustering energy optimization (EPSO-CEO) system that minimizes the use of power in each node by using centralized clusters and optimized cluster heads. Nevertheless, none of these studies jointly address the lifetime and fault-tolerance aspects in their routing algorithms with a convergent model.

The authors in Reference 13 present a model to prevent the unnecessary convergence of crowd by setting upper and lower search space bounds so as to enable the crowd to find solutions for diverse applications. Furthermore, the authors in Reference 13 demonstrate the performance of a load distribution system so as to address the optimization problem and facilitate prime network selection. Moreover, the authors account for the network bandwidth and the errors in the ideal data delivery system in different networks with reduced cost. To improve the lifetime and increase the bandwidth of the energy-proficient distributed clustering practice, the authors in Reference 14 employ the use of energy-aware, delay-tolerant, and centralized approaches.

6.3 System Model

The routing method that has been projected in this research uses a fault-tolerant system in two-tiered heterogeneous WSNs that comprise super/smart nodes that have plenty of resources and simple/light sensor nodes with limited battery capacity and absolute QoS limitations. Nevertheless, to get a more resilient fault-tolerant network model, we look for a k-disjoint multipath routing approach. Furthermore, we look at a many-to-one traffic system, in which super nodes and common nodes connect with the proper degree. Below, we list some important explanations of some terms before introducing the system model.

For every disjoined/isolated node, we have to use it to build a k-disjoint multipath route and increase the number of marginal paths, rendering a fault-tolerant network. Our model is based on the assumption that a given node can connect or disconnect with some nodes that are not among those that are on the k-disjoint multipath between the node and a super node. In this study, node-disjointness relations are modeled as a directed graph $G(V,E)$, where $|V| = \{v_1, v_2, ..., v_N, v_{N+1}, ..., v_{N+M}\}$ is the fixed number of nodes or particles, N indicates the sensor node while M signifies super/smart nodes; G represents the set of paths and the relationship between a pair of super nodes, and a pair of particles is the number of edges E in G. $E = \{(v_i, v_j) \mid Hop(v_i, v_j) \leq \tau\}$, where $Hop(v_i, v_j)$ is the distance between v_i and v_j. $P(v_i, v_j)$ is a path that runs from v_i to v_j in graph G. It is a sequence of edges we get when we go from v_i to v_j, where $i = j = 1, 2, ..., N + M$.

Hence, we can describe G as a set of unconventional routes $p_i(v_i, v_j)$. $e \in p_i(v_i, v_j)$, (v_N, v_{N+M}) denotes a connection between any two nodes, $E(v_i, v_j \in p_i(v_i, v_j))$ is the node disjoint between $p_i(v_i, v_j)$, (v_N, v_{N+M}), and e. Hence, we can obtain k-disjoint paths in G. We use the amount of energy consumed by a multipath, delay, and throughput to evaluate how best a multipath performs. We use the roots in References 13 and 14 to come up with a solution of the objective function that minimizes the energy consumption and average delay and maximizes the system throughput and network lifetime.

6.3.1 Problem Formulation

For our problem statement, we are looking to design a k-disjoint multipath for a fault-tolerant system, which uses a UAV to transmit multimedia to a super node located in a two-tiered WSN. The model is constructed in such a way that each sensor node in the network is within the transmission range of each other node. This helps in minimizing QoS parameters such as the transmission power level and the average delay while maintaining a k-disjoint multipath route. In this system, every sensor node is connected to at least one super node with a k-disjoint multipath. Accordingly, a k-disjoint multipath is constructed by connecting a group of super nodes with a bunch of sensor nodes that can modify their transmission range to a prime value. The transmission range of each sensor should be such that the minimum amount of energy is used, while still maintaining a k-disjoint multipath, and all the parameters for QoS are still upheld.

6.3.2 Energy Model

For proper energy limitations, we need to consider the number of hops and the distance between two UAVs along the predefined path. The neighborhood topology mentioned in Section 6.3 is used. Each sensor node is within the transmission range of its neighbor sensor node. Given that the transmission range is equal to τ (>0), then the neighborhood is formulated by: $\aleph_{u,v} = \{v, u \neq v \| n_u - n_v \| \leq t_u\}$. It is worth recognizing that there is a chance that this might change during the dynamic network lifetime. Moreover, unless all constrains are met, there will be a division in the multipath and the neighborhood will be reconstructed. We can use the constrains to change the topology of the system, which will eventually lead to solving the optimal power problem. The authors in Reference 13 use the method of cutoff value to determine the lower and upper bounds of the number of hops and the transmission range. E_{elec} represents the energy that is deprived when using the transmitter and receiver circuitry. The energy used by the receiver to obtain a proper signal-to-noise ratio is represented by ε_{mp}. The amount of energy loss during transmission is α, and $\tau_{n(nu, nv)}$ is the transmission range. To conclude, the function used to get the minimum amount of energy used in one node to transport data of length L_p for a distant of τ is formulated by

$$min \ \vec{Z} \tag{6.1}$$

so that

$$hop = \sqrt[\alpha]{\tau_n(n_u,n_v)\left(\frac{3\varepsilon_{mp}}{2E_{elec_n(n_u,n_v)}}\right)} \leq \tau_n(n_u,n_v) \tag{6.2}$$

and

$$\vec{Z} = Energy_{nd} = L_p\left\{\sum_{n_n}^{n_d} 2\left[E_{elec_{nd}} + \varepsilon_{mp}(n_{sd})^\alpha\right]\right\}. \tag{6.3}$$

The above energy value for the selected path can change according to the selected upper/lower bounds E_{min}, and E_{max}, respectively. It represents the minimum and the maximum constants.

6.3.3 Delay Model

In this research, we consider the delay definition that depends on the hop count, denoted as $\varphi(\xi_i,\xi_j)$. φ represents the delay between two nodes; its definition is determined by the ideal number of hops. Given the optimal number of hops in Equation 6.3, which represents the minimum delay between two nodes, we can formulate and optimize the route selection while considering delay and network resource constraints. This optimization problem should consider both source and intermediate nodes periodically in the immediate neighborhood. Additionally, if one sensor node gratifies one QoS, the problem converges and all QoS requirements will be achieved. Consequently, the end-to-end delay for a given path P between is ξ_{Source} and ξ_{Sink} is described as

$$\varphi_{sourcessSink}(L_p) = min\left\{\sum_{\xi_i}\varphi(\xi_i,\xi_j)\right\}, \tag{6.4}$$

where $\varphi_{SourcessSink}$ denotes the minimum delay that we can achieve when we send data through paths between ξ_{Source} and ξ_{Sink}. This time consists of the time for transmission, retransmission, staying idle, queuing, propagation, and processing. And thus, considering

$$\sum_{v=1}\varphi(\xi_i,\xi_j) \leq X_v\Delta_\varphi, \tag{6.5}$$

the average delay per sensor node is equal to ξ. Assuming the hop count on a path between ξ_{Source} and ξ_{Sink} is given by η_{ij}, and the delay along this path is L_e^{φ}. The hop delay constraint can be signified by $L_e^{\varphi} = (\Delta_{\varphi} - \varphi^e / \eta_{ij})$. Accordingly, we can rewrite the constraint in Equation 6.5 as $\sum_{u=1} \varphi(\xi_i, \xi_j) \leq X_u L_e^{\varphi}$.

6.3.4 Throughput Model

According to Reference 14, a definition of throughput can be used to represent the number of data packets successfully transmitted. This can help in calculating the optimal hop count while maximizing the network throughput. This throughput is computed via the following: $Th = \left(L_e^{\varphi} / \xi \right) * TR$, where TR is the transmission rate.

6.4 Particle Swarm Optimization (PSO) Algorithm

Every sensor/particle is allocated to a k-disjoint multipath according to the sensor transmission level and required separation distance per hop, as alluded to in Equation 6.3. This is performed while considering numerous swarm particles' attributes. In this approach, sensor nodes have the capacity to enhance agreeable learning conduct by trading path-related messages with their neighbors. After trading/exchanging these messages, every node/particle figures the disjoint ways and expands the neighborhood set as alluded to in Equation 6.3. As indicated by Equations 6.2 and 6.4, another potential set of paths will be formed and prioritized, and thus, hops per the k-disjoint multipath will be adaptively changed according to the particle speed $v_{(i,j)}$ that is refreshed after each iteration to fulfill the desired QoS requirements.

Given that a k-disjoint multipath can have m descriptive QoS attributes, the position and velocity of the particle v is given by an m dimensional vector $|V| = \{v_1, v_2, \ldots, v_N, v_{N+1}, \ldots, v_{N+M}\}$. The proposed swarm algorithm in this chapter contains p_{best} and g_{best}, which are the personal and global best positions, respectively. By solving Equations 6.2 and 6.4 in terms of the average amount of consumed energy and average delay, we find the nodes that connect the entire searching space in every iteration. Trading control messages between the nodes can further trigger them, and they are then defined as extreme value and global extreme value. Consequently, this leads to the extreme value within the feasible search space toward which we progress after every iteration. Hence, the nodes that tend to diverge are excluded. The personal best position of the swarm is brought about by the dissemination of good objective functions as denoted in Equations 6.2 and 6.4. These equations are concerned with the information exchange, while satisfying the constrains that are used to get the velocity and then find the ideal multipath route, as stated before. In every path, the personal best position of a particle $v_{(i,j)}$ is given by $p_{best,v(i,j)} = (p_{(best,v1),(best,v2),\ \ldots\ ,(best,vN),(best,vN+M)})$; similarly, the global-best position of a particle is given by $g_{best,v(i,j)} = (g_{(best,v1),(best,v2),\ \ldots\ ,(best,vN),(best,vN+M)})$. The extent to which the $p_{best,v(i,j)}$ affects the equation is given by the coefficient of constraints φ_1; similarly, the effects of the global-best position are denoted by the coefficient of constraints φ_2. The velocity of the updated function drives what we call the CPSO, which can be defined mathematically as follows:

$$\vec{v} := x.v_u, \tag{6.6}$$

$$\vec{v} := x. + Z\left(0,\emptyset_1\right) \otimes \left(\overrightarrow{p_v} - \overrightarrow{x_u}\right), \tag{6.7}$$

and

$$\vec{v} := x. + Z\left(0,\emptyset_2\right) \otimes \left(\overrightarrow{g_v} - \overrightarrow{x_u}\right), \tag{6.8}$$

where $\vec{Z}$ represents the distribution of the objective function, found after satisfying the constrains mentioned above. χ is the constriction coefficient that assists in balancing global and local probes. It is represented as $x = 2 / \emptyset + \sqrt{\emptyset^2 - 4\emptyset}$, *where* $\emptyset = \emptyset_1 + \emptyset_2 > 1$. Equation 6.6 is defined as the velocity update function and regarded as the momentum function, which gives the particle's/node's present direction. Equation 6.7 is called social component; it has the ability of being drawn toward the best solutions as assessed by the neighbors. Equation 6.8 represents the cognitive module, with the ability of being drawn toward earlier results that symbolize the node behavior. The only difference between CPSO and the fully particle multipath swarm optimization (FPMSO) algorithm is the function used to update the particle velocity. This means that we take into account the best position not only of the node but also of all its neighbors. Hence, the velocity update function can be formulated as

$$\vec{v}_v = x \left(\vec{v}_v + \frac{1}{K_v} \sum_{m=1}^{k_v} \vec{Z}(0,\emptyset_1) \otimes (\vec{p}_v - \vec{x}_v) \right). \tag{6.9}$$

We can ignore some of the node fault-tolerance messages that can lead to trapping in local optimal solutions by eliminating the exchanged information about the personal best $p_{best,v(i,j)}$. Consequently, this can lead to the increase of the node's ability to learn from the experience of other nodes. Hence, the performance of the algorithm highly rests on influence of the nodes while satisfying the objective function. Algorithm 6.1 denotes the pseudocode of the proposed CPSO algorithm. It finds the p_{best}'s objective function given by Equations 6.2 and 6.4, first in terms of the consumed energy and average delay and then in terms of finding the least value of objective function in the p_{best}'s objective function for the *k*-disjoint multipath. Then, it assists in avoiding velocity fit and computes the constriction value χ, as shown in $\vec{v}_v$. It updates the velocity value and finally establishes a better fault-tolerant multipath route from which the optimal nodes are chosen.

Algorithm 6.1: CPSO

---

1. input: Objective functions $f(x)$

2. $X := \{x_1, \ldots, x_n\} := InitParticle\left(\overrightarrow{lb}, \overrightarrow{ub}\right) \rightarrow \forall_p \in \{1, \ldots, n\} : \vec{x}_n := \vec{U}\left(\overrightarrow{lb}, \overrightarrow{ub}\right)$

3. $V := \{v_1, \ldots, v_n\} := InitParticleVelocities\left(\overrightarrow{lb}, \overrightarrow{ub}\right) \rightarrow \forall_p \in \{1, \ldots, n\} :$

$$\vec{v}_n := \left(\overrightarrow{lb} - \overrightarrow{ub}\right) \otimes \vec{U}(0,1) - \frac{1}{2}\left(\overrightarrow{ub}, \overrightarrow{lb}\right)$$

4. $Y := \{\vec{y}_1, \ldots, \vec{y}_n\} := EvaluateObjectfunction(X) \rightarrow \forall_p \in \{1, \ldots, n\} : y_n := f\overrightarrow{x_p}$

5. $P := \{\vec{p}_1, \ldots, \vec{p}_n\} := Initllocallocallyoptimal(X) \rightarrow X$

6. $P := \{p_1^f, \ldots, p_n^f\} := InitObjeectivefunction(Y) \rightarrow Y$

7. $G := \{\vec{g}_1, \ldots, \vec{g}_n\} := Initgloballyoptimal(P, T) \rightarrow P$

8. $G := \{g_1^f, \ldots, g_n^f\} := Initgloballyoptimal(P^f, T) \rightarrow P^f$

9. **while** termination condition nor met do **do**

10.   **for** each particle node p of n do **do**

11. $u_p = x * \begin{pmatrix} \overrightarrow{u_p} + \vec{\psi}(0, \varphi_2) \otimes \left(\overrightarrow{locallyoptimal}_p - \overrightarrow{x_p}\right) \\ + \vec{\psi}(0, \varphi_2) \otimes \left(\overrightarrow{locallyoptimal}_p - \overrightarrow{x_p}\right) \end{pmatrix}$

12. $\overrightarrow{x_p} := \overrightarrow{x_p} + \overrightarrow{v_p}$

13.   **end for**

14. $Y := EvaluateObjectivefunction(X, f)$

15. $P, P^f := Updatelocallyoptimal(X, Y) \rightarrow \forall_p \in \{1, \ldots, n\} : \overrightarrow{p_p}, p_p^f$

$$:= \begin{cases} \overrightarrow{x_p}, y_i & \text{if } y_i \text{ better than } p_p^f \\ \overrightarrow{p_p}, p_p^f & \text{otherwise} \end{cases}$$

16. $G, G^f := Updategloballyoptimal(P, P^f, T) \rightarrow \forall_p \in \{1, \ldots, n\} :$

$$\overrightarrow{g_p}, g_p^f := best\left(P_{T_p}, P_{T_p}^f\right), \text{where } T_p \text{ are the neighbors of } p$$

17. **End while**

Table 6.1 Assumed Parameters

Parameter	Value
Message payload	64 bytes
Data length p	2000 bits
Transmission range	12.00 m
Tx data rate	250 kbps
Eelec	50 nJ/bit
Total number of UAVs	50 sensor nodes
εmp	0.0013 pJ/bitm^2
Topology structure	Square (1000 m * 1000 m)
εfs	10 pJ/bitm^2

6.5 Performance Evaluation

Keeping in mind the end goal of evaluating the execution of the proposed swarm method, we perform broad reenactments. We have implemented the aforementioned algorithms (the Particle Multi-Swarm Optimization [PMSO], the CPSO, and the FPMSO) by using MATLAB in order to evaluate their objective functions and visualize their outputs. We use 100 sensors and 50 UAVs dispersed uniformly in $1000 \times 1000 \times 100$ m^3 deployment space. The path loss exponent for the wireless communication model is chosen to be 2. The underlying estimation of the sensors' transmission range is set to be 100 m in order to assure the association among UAVs and sensor nodes while fulfilling the pertinent QoS requirements. Further simulation parameters are compacted in Table 6.1.

6.5.1 Simulation Results

Figure 6.1 presents the total energy consumption in the assumed topology with a maximum path length of 5 hops. We remark that the aggregate energy utilization in the k-disjoint multipath created by the proposed PMSO approach is superior to CPSO because settling the objective function used by CPSO experiences issues in finding a k-disjoint multipath after recouping from operational failures in the immense search space. Furthermore, since this results in not being able to substitute the arranged multipath with some other options, more energy usage is experienced.

Another critical comment that is identified by traded messages for adaptation to failure between the super nodes (UAVs) and sensors is that CPSO performs essentially worse than both FPMSO and CMPSO. This is because CPSO necessitates that fundamentally more control packets trade among the neighbors. In this way,

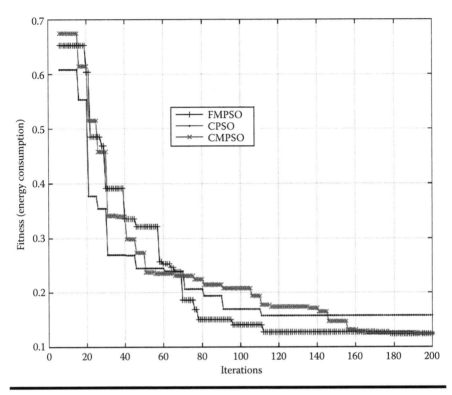

Figure 6.1   The swarm optimization routing versions vs. energy consumption.

CPSO needs to discover a *k*-disjoint multipath in its nearby neighborhood, even though the FPMSO and CMPSO can straightforwardly scan for routes utilizing fewer control packets among the nearby hops. Subsequently, the *k*-disjoint multipath for FPMSO and CMPSO can bring down the aggregate energy usage when contrasted with the CPSO approach.

Figure 6.2 demonstrates the average deferral of the chosen ideal *k*-disjoint multipath from the sender to the destination. We can watch that the assessed approaches, FPMSO and CMPSO, have exhibited a lower delay for every hop when contrasted with CPSO. This can come back to the determination and upkeep of a *k*-disjoint multipath for adaptation to failure that can fulfill the hop necessities by choosing the following bounce in the area of every hop. Therefore, it necessitates fundamentally fewer control messages for adaptation to noncritical failure when contrasted with CPSO for choosing and keeping up a 1-hop neighborhood. Hence, we can say that both CMPSO and FPMSO are more practical than CPSO.

Throughput might vary because of a high-bit error rate (BER) and other surrounding changing conditions outdoors. Accordingly, we show in Figure 6.3 the impact of tackling the objective function alluded to in Equation 6.4. We notice that, while expanding the ideal number of hops and trying to minimize the average

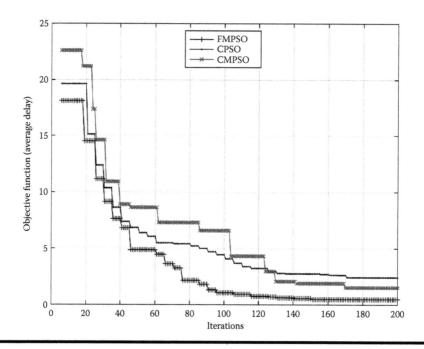

Figure 6.2   The swarm optimization routing versions vs. average delay.

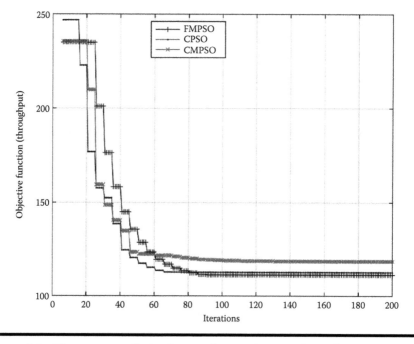

Figure 6.3   The swarm optimization routing versions vs. throughput.

delay, throughput degrades. This is a normal result since limiting latency under the previously mentioned requirements due to the reduced number of traded control messages. Therefore, the most practical hops on the route can be acquired effectively.

6.6 Conclusion

In this research, we offer a bio-inspired swarm algorithm that constructs, recovers, and finds k-disjoint multipath routes in a network of sensors and UAVs. Two information positions, namely the personal-best position and the global position, are considered in the form of velocity updates to enhance the performance of the IoT network. In order to validate this algorithm, we assessed multiple objective functions that consider throughput, average energy consumption, and average end-to-end delay. Our results show that using the characteristics of all personal-best information is a valid strategy for the purposes of improving the PMSO performance. Moreover, the proposed algorithm has also been compared with similar algorithms that optimize the energy consumption and average delay on the explored paths toward the destination.

For the future, we see great potential and need to study various aspects of 5G/IoT integration with the existing sensor network architectures in different levels for more successful industrial applications. The popularity of 5G, the problem of slicing Internet traffic, and the fact that a significant slice is expected to be reserved for sensory applications encourages further attempts in this domain.

References

1. I. Bekmezci, O.K. Sahingoz, and S. Temel, Flying ad-hoc networks (FANETs): A survey, *Ad Hoc Networks*, vol. 11, no. 3, pp. 1254–1270, 2013.
2. V. Petrov et al., When IoT keeps people in the loop: A path towards a new global utility, *arXiv preprint arXiv:1703.00541*, 2017.
3. A. Hadjidj, A. Bouabdallah, and Y. Challal, HDMRP: An efficient fault-tolerant multipath routing protocol for heterogeneous wireless sensor networks, in *Quality, Reliability, Security and Robustness in Heterogeneous Networks. QShine 2010. Lecture Notes of the Institute for Computer Sciences, Social Informatics and Telecommunications Engineering*, X. Zhang and D. Qiao, Eds, vol. 74. Springer: Berlin, Heidelberg.
4. S. Jiang et al., Linear decision fusion under the control of constrained PSO for WSNs, *International Journal of Distributed Sensor Networks*, vol. 8, no. 1, pp. 871596, 2012.
5. C.-H. Wu and Y.-C. Chung, Heterogeneous wireless sensor network deployment and topology control based on irregular sensor model, in *Proceedings in Advances in Grid and Pervasive Computing*, C. Cérin and K.-C. Li, Eds, Springer: Berlin, pp. 78–88, 2007.
6. J.K. Vis, *Particle Swarm Optimizer for Finding Robust Optima*, Leiden, The Netherlands, http://www.liacs.nl/assets/Bachelorscripties/2009-12JonathanVis.pdf, January 15, 2015.

7. W.H. Lim and N.A. Mat Isa, Particle swarm optimization with adaptive time-vary-ing topology connectivity, *Applied Soft Computing*, vol. 24, pp. 623–642, 2014.

8. M.B.K. Dhir, A survey on fault tolerant multipath routing protocols in wireless sensor networks, *Global Journal of Computer Science and Technology*, vol. 15, no. 3, pp. 1–13, 2016.

9. M.A. Adnan, M.A. Razzaque, I. Ahmed, and I.F. Isnin, Bio-Mimic optimization strategies in wireless sensor networks: A survey, *Sensors*, vol. 14, no. 1, pp. 299–345, 2014.

10. Y.H. Robinson and M. Rajaram, Energy-aware multipath routing scheme based on particle swarm optimization in mobile ad hoc networks. *The Scientific World Journal*, vol. 2015, p. 9, 2015.

11. M. Azharuddin and P.K. Jana, A PSO based fault yolerant routing algorithm for wireless sensor networks, in *Information Systems Design and Intelligent Applications: Proceedings of Second International Conference INDIA 2015*, vol. 1, J.K. Mandal et al., Eds, Springer: New Delhi, pp. 329–336, 2015.

12. C. Vimalarani, R. Subramanian, and S.N. Sivanandam, An enhanced PSO-based clustering energy optimization algorithm for wireless sensor network, *The Scientific World Journal*, vol. 2016, p. 11, 2016.

13. M.Z. Hasan, F. Al-Turjman, and H. Al-Rizzo, Optimized multi- constrained quality-of-service multipath routing approach for multimedia sensor networks, *IEEE Sensors Journal*, vol. 17, no. 7, pp. 2298–2309, 2017.

14. M.Z. Hasan, F. Al-Turjman, and H. Al-Rizzo, Evaluation of a dutycycled proto-col for TDMA-based Wireless Sensor Networks, *International Conference in Wireless Communications and Mobile Computing (IWCMC)*, pp. 964–969, 2016.

Chapter 7

Evaluation of a Duty-Cycled Asynchronous X-MAC Protocol for VSNs

7.1 Introduction

Vehicular sensor networks (VSNs) allow limited-range sensor devices to communicate with each other [1]. VSNs are promising solutions for specific cases within the Internet of things (IoT), which allow the integration of different objects to communicate with each other in dynamic environments [2]. The current trends in VSNs allow different deployment architectures for vehicular networks in highways and in urban and rural environments to support many applications with different quality-of-service (QoS) requirements [3]. Basically, VSNs came to allow communication among nearby vehicles as well as with fixed roadside equipment that leads to three different configurations: vehicle-to-vehicle (V2V), vehicle-to-infrastructure (V2I), and hybrid network architectures, as illustrated in Figure 7.1 [4]. Features of these configurations encounter new challenges in order to expand from being a network of computers to a network of both computers and things.

Devices in the IoT connect with each other using a variety of protocols, and there still exists a large number of devices that use older communication protocols but have diverse real-time needs. Therefore, VSNs offer integrated communication

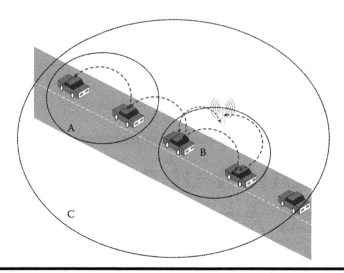

Figure 7.1 Basic architecture of VSNs: (A) vehicle-to-vehicle (V2V), (B) vehicle-to-infrastructure (V2I), (C) hybrid.

protocols for effectively monitoring the physical world, especially in urban areas where a high concentration of vehicles equipped with onboard sensors is expected [5,6]. Despite this, integration has benefits such as increasing revenue, reducing costs, and energy efficiency. However, there exists a serious problem with traffic congestion in decision-making for vehicular traffic, which is a challenge due to particular characteristics, such as the highly dynamic topology and the intermittent connectivity [4]. Consequently, VSNs have challenges in supporting real-time traffic information that can significantly improve the safety of transportation and reduce the traffic congestion [7]. This information will help drivers to make smarter decisions in a timely manner to prevent accidents, improve the efficiency of the selected route, and provide a safer distance among other vehicles. Therefore, the duty of the embedded sensor is to capture images and measure distance all around a vehicle in order to monitor traffic in an allocated area, while utilizing different devices that can measure several physical traffic parameters [5]. Hence, the view of the vehicle as a sensor platform can improve the traffic flow via supporting communication with roadside infrastructure in order to provide ubiquitous coverage [8]. The relative velocities of vehicles are much higher than 50 km/h in urban environments and more than 100 km/h on the highway [4]. Vehicles also move in different directions. Thus, vehicles can quickly access or leave the network in a very short period of time. This results in more frequent changes in the network topology, which affects the network design significantly. For example, the routing protocol design will be more difficult due to hidden/exposed terminal problems in medium access control (MAC) protocols [9]. Vehicles are typically not affected by strict energy constraints and can be easily equipped with sensor platforms [10].

Meanwhile, VSNs represent a significantly novel and challenging deployment scenario, which differs considerably from the traditional wireless sensor network (WSN) and thus requires innovative solutions in the MAC layer [11]. However, designing an integrated architecture for both WSNs and VSNs often starts with the definition of a MAC protocol since it is a fundamental issue in determining the energy consumption properties and the basic data transport capabilities of the network [12].

This design of an efficient and effective sensory MAC protocol providing QoS requirements for real-time traffic management is considered as the most important step in end-to-end QoS provisioning over VSNs [13] since it regulates nodes' access to a shared channel and has become a major active research area in recent years [14]. This regulation explicates as a duty cycling approach that is considered as one of the primary mechanisms for providing QoS in VSNs [15].

Particularly, duty cycling means that every node in the network is periodically alternating between an awake and a sleep state [16]. Therefore, the duration of a duty cycle is equivalent to the time of an awake state plus the time of a sleep state [17]. The idle state has been founded in the IEEE 802.11p standard for vehicular communications that consume substantial energy to transmit up to 1000 messages with 32 dBm and therefore should be avoided in VSNs [4].

To understand the performance of VSNs and in order to optimize the designed routing protocol [18], an accurate analytical framework for the MAC protocol is required. The main idea of this framework is to provide an analytical scheme that dynamically adapts the vehicles' rate of transmission according to their priority. The analytical model should describe the effects of assigning various values, including the density and transmission range of vehicles to protocol parameters under specific, given scenarios in order to achieve the QoS requirements.

The remainder of the chapter is organized as follows. Section 7.2 discusses some related works devoted entirely to analytical modeling of MAC protocols. Section 7.3 introduces an overview of the hidden terminal problem and accordingly provides the power consumption, delay, and network throughput analysis. Section 7.4 introduces the behavior of the X-MAC protocol under specific network conditions through using the proposed Markov model. Meanwhile, Section 7.5 introduces the performance analysis of the synchronized X-MAC protocol. Furthermore, Section 7.6 introduces detailed simulation and analysis for the performance of a synchronized X-MAC protocol of various scenarios. Finally, Section 7.7 concludes the work in this chapter.

7.2 Related Works

To better understand the mechanism of a MAC protocol, it is useful to realize that MAC protocols consists usually of three main logical components [19]. First, a collision avoidance (CA) algorithm uses physical carrier sensing to register and/or reserve the channel for the duration of the data transmission. Second, a contention resolution algorithm uses mechanisms such as back-off to regulate the access to the channel [19]. Third, there is a distributed coordination function (DCF) that is not specifically designed for high-mobility network [7].

Various MAC protocols have been proposed to mitigate the adverse effects of hidden terminals through CA, since the hidden problem has demonstrated its energy-saving capabilities [11]. However, in a heterogeneous wireless network, a hidden problem should be defined as a node out of the range of the sender that covers the receiver. Most CA algorithms are based on sender initiation, including an exchange of short request-to-send (RTS) and clear-to-send (CTS) messages between a pair of sending and receiving nodes before the transmissions of the actual data packet and the optional acknowledgment packet [8]. In receiver initiation, a receiver broadcasts a probing packet whenever it wakes up from its sleeping state, while a sender with a data packet to transmit waits in the listening state until the probing packets from the receiver is received. Therefore, the receiver-initiated MAC protocol degrades the network performance with asymmetric links due to several experienced sender failures in receiving the probing packets from the receiver. Hence, the asymmetric links waste energy, increase delay, and degrade the packet received ratio (PRR). Meanwhile, RTS and CTS message-exchange mechanisms cannot be the solution for VSNs since these exchange messages may not be able to arrive to all hidden nodes [20].

MAC protocols can be divided into two main categories of duty-cycled MAC protocols [17]. One is synchronized protocols, like S-MAC [21] and T-MAC [22]. The other is asynchronous protocols, like X-MAC [17] and B-MAC [23]. Asynchronous duty-cycled MAC protocols remove the energy overhead for synchronization and are easier to implement as they do not require local synchronization [24]. The X-MAC protocol uses data packets as preambles and adapts it for sparse networks as the energy and collision increase linearly with the node density. Thus, the performance of the X-MAC protocol is evaluated in this work when equipped with a CA algorithm to address the performance degradation of wireless multi-hop communications with hidden terminals and their impact on MAC protocols. Additionally, the X-MAC contention-based protocol does not perform well in asymmetric scenarios due to the quality of the link from the receiver to the sender, which causes the hidden terminal problems and can be avoided if the communication was not receiver-initiated [25].

Many researchers have evaluated the performance of various network protocols through simulations [26]. However, the simulation environment/software is usually too expensive and/or time-consuming, especially while considering a huge network size/capacity [27]. Meanwhile, the analytical models can be more effective in such cases, since the scale of modern networks and the degree of complexity often necessitate the use of simplified assumptions, (for example, Markov, Poisson traffic, or other models). Furthermore, it is hard to capture the dynamic nature of a network without an analytical model; therefore, an analytical model is needed to provide insight into the performance of both routing and MAC protocols [27].

Bianchi [28] proposed an accurate analytical model to analyze the performance of single-hop IEEE 802.11. Ziouva [28] improved Bianchi's model by adding a deriving saturation delay beside throughput. In an area other than IEEE 80.11, specifically in WSNs, the authors in Reference 29 proposed a radio model to compute the lower bound of the X-MAC protocol. A new hybrid MAC scheme called zebra MAC (ZMAC), which combines the strengths of time division multiple access (TDMA) and code division multiple access (CDMA) while offsetting their weaknesses, is proposed in Reference 30 for sensor networks. The authors in Reference 31 implement an efficient TDMA protocol that applies a duty-cycling function for multi-hop WSNs using a semi-Markov chain. The authors try to avoid channel access problems such as overhearing and hidden terminals by adapting the wake-up/sleep state of each node to the actual operational conditions such as traffic demand and node density. Short-range V2V communication was investigated in Reference 32. The authors present a study in which effective information such as the message size, transmission range, and velocity of vehicles is exchanged. Such exchanged parameters are considered as factors in analytical models to evaluate the performance of communication. Meanwhile, the authors in Reference 33 present a study of connectivity in vehicular ad hoc networks in traffic free-flow. Actually, the authors use the analytical model to describe the distribution of distances between the vehicles, the traffic flow on the highway in evaluating the effects of various systems' parameters (such as distribution of velocity, traffic flow, and transmission range of vehicles), and the network on connectivity.

B-MAC [23] considers the default MAC protocol for Mica2 that allows an application to implement its own MAC through a well-defined interface. To achieve channel utilization and low-power operations, the authors adopt low-power listening (LPL) and scheduling of the clear channel sensing access (CCA) technique to reduce duty cycle and minimize idle listening. Yang, in Reference 34, modeled and analyzed the throughput of a synchronized duty-cycle S-MAC protocol for WSNs. The S-MAC protocol has different rules for accessing the media as compared against other MAC protocols such as X-MAC or B-MAC. Yang also proposed a Markov model to analyze the throughput of X-MAC. It should be emphasized that our proposed model is fundamentally different from the one proposed by Yang [29]. Our proposed model analyzes the performance of the X-MAC protocol when it is equipped with the CA algorithm and aimed at addressing the performance degradation of VSNs with a hidden terminal. However, none included the hidden terminal problem in their analytical models. The paper focuses on the evaluate of an adaptive energy-efficient X-MAC protocol for duty-cycled VSNs. A Markov queuing model is proposed for modeling the behavior of the X-MAC contention based on the specific sleep/wake-up pattern in the duty-cycle. Our proposed model quantifies the desirable QoS metrics for contention-based MAC protocols in multi-hop fashion to address the hidden terminal problem and provide fairness in medium sharing among the vehicles.

7.3 Overview of the X-MAC Protocol

Asynchronous protocols have promising applications in WSNs because they avoid synchronization overhead and hence provide higher energy efficiency than synchronized MAC protocols [29]. Many variations of asynchronous duty-cycled MAC protocols have been proposed to improve energy efficiency and packet latency by allowing each node to independently and periodically sleep to save energy [24]. Additionally, asynchronous protocols use a series of short preamble packets to avoid synchronous overheads and hence have higher energy efficiency than synchronized MAC protocols [12]. These short preamble packets carry the address information of the sink node. As a result, the intermediate nodes can go to sleep as soon as they hear the first short preamble. Moreover, the sink can reply with an Acknowledgment (ACK) message in between two successive short preambles to stop the timeline and to start transfering the data packets [17].

Figure 7.2 indicates the LPL timelines of short preamble packets of the X-MAC protocol that effectively constitute a single long preamble [17]. Whereas a node wakes up periodically to send and receive a short preamble packet to the sink

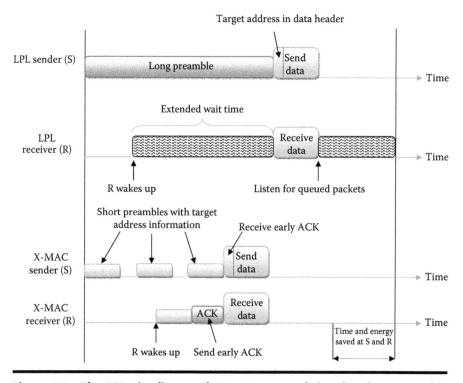

Figure 7.2 The LPL timelines and X-MAC protocol for the short preamble approach. (Adapted from N. Saxena et al., *Computer Networks*, vol. 52, pp. 2532–2542, 2008.)

node (which is still sleeping), it looks at the identification (ID) of the sink node included in the packet. If the other nodes are not the intended recipient, they go to sleep quickly to avoid the overhearing problem; otherwise, they remain awake for the subsequent packet. Hence, the interval between two successive wake-up periods is called a cycle (denoted as T) for every node. Every node begins its duty with a fixed cycle length determined with an arbitrary offset and continues its duty cycling as if the medium had been idle. For each successfully delivered data packet, the average communication time is $(T/2)$ pulsing the length of the data packet L_{DATA} [29].

X-MAC also has collisions related to increasing the network density; that is, the number of senders increases and they wake up and begin to send their preamble at the same time. Thus, all nodes including sink nodes cannot determine the destination address information in the preamble when collision occurs among nodes. In this case, the sender continues sending preambles until the next wake-up time [17]. Hence, for each colliding data packet, the average communication of sending a data packet is also T.

This chapter proposes a mathematical model for the X-MAC protocol that includes the effects of a CA algorithm. Our proposed model focuses onto two main contributions, which are (1) solving the problems of medium contention, such as hidden or exposed terminal problems, and (2) providing resource reservation for real-time traffic control systems in a distributed vehicle-based sensor environment. Moreover, supporting QoS in the routing or transport layer cannot be provided unless the assumption of MAC protocols solves the problems of medium contention and supports reliable communication [35].

Our proposed model acts as an analyzer for the performance of X-MAC since the Markov model is used to describe the behavior of accessing of synchronized duty nodes to the channel. The proposed model elaborates on which type of low duty-cycled MAC protocols should be selected in order to resource the wireless channel reservation that assures the desirable QoS level in real-time traffic control systems.

7.4 Markov Model of X-MAC

We propose a Markov model to describe the behavior of X-MAC and investigate the QoS parameters under various network and channel conditions. However, to estimate the effect of the hidden problem, it is necessary to examine the transmission among one-hop neighbor nodes at regions where possible hidden problems occur as depicted in Figure 7.3 in which node A transmits a frame to node B and node D transmits a frame to node B. Thus, node D will be able to transmit because it is unable to detect the transmission of node A or node C. This means that both nodes C and D are hidden from each other, resulting in a collision at node B that causes a serious QoS degradation, especially in high-data-rate sensor applications [29,36].

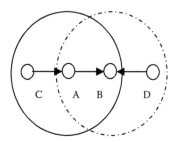

Figure 7.3   Hidden terminal problem in contention-based protocols.

In order to satisfy the QoS requirements for real-time traffic control systems, a Markov discrete-time stochastic process M/M/1 queuing model is proposed under realistic conditions for duty cycling with a schedule-driven operation in VSNs. We consider a simple traffic model for vehicles moving along a straight road with random velocity. The arrival and departure of the data packets are regulated under a realistic assumption of a finite queue size. Therefore, the proposed model makes the following assumptions:

1. Event arrivals denote a stochastic process $\{A(t)|t \geq 0\}$ that represents the total number of arrivals that have occurred from time 0 to time t; this procedure creates an independent Poisson process at each node, and the number of packet arrivals in any time slot is distributed with a Poisson process with parameter $\lambda\tau$, for time of arrival $t\,\tau \geq 0$.
2. Let $\pi_i(t)$ denote the steady state of power for the node at time t; the interarrival $\delta \geq 0$ times (that is, the distribution of time at state $\imath$ before marking the transition) are independent and exponentially distributed with the λ, where $o(\delta)$ is defined as a function of δ such that $\lim_{\delta \to 0} \frac{o(\delta)}{\delta}$.
3. The queueing discipline of data packets is first come, first served (FCFS).
4. The queueing system assumes equilibrium under the condition that the probability of arrival is less than the independent probability of transmitting the information packet, or $\lambda < \beta$.
5. The processing and radio transmission times are independent and identical (*i.i.d.*) with an arbitrary distribution.
6. Retransmission is supported.
7. When an event is sensed, the node processes it and sends the information packet with a probability of transmission per node per cycle, and every sensor node in the network has an independent probability of transmitting information packet β in the duty cycle.

These assumptions, which have been verified as valid approximations of realistic scenarios, are made based on Reference 37. The proposed Markov model shows that the power transition of each sensor node in the network may be modeled by a discrete-time M/M/1 Markov chain, which represents a different predefined status for a node for an event at the wake-up/sleep mode of the duty cycle. The proposed model considers the following assumptions:

a. The vehicles are equipped with sensor nodes in a network and are assumed to be two-dimensionally Poisson-distributed over a domain with density ρ. Therefore, the probability of finding n neighboring nodes in an area S is given by [34]

$$p(n, S) = \frac{(\rho S)^n}{n!} \exp^{(-\rho S)}.$$

$$(7.1)$$

b. If each node has the same transmission range T_R for transmission and receiving in time-slotted mode, then the number of n neighbors within a circular region of T_R is given as [38]

$$n = \rho \pi T_R.$$

$$(7.2)$$

c. All nodes are assumed to be saturated, that is, they always have some data packet in the queue waiting for transmission, and all data packets have the same length.

d. The time step value T is assumed not only for the propagation delay but also for the transmission delays, processing delay, carrier sense delay, and queuing delay.

e. Each sensor node has the transmission range T_R defined as [36]

$$T_R = \sqrt[\beta]{\frac{G_t G_t \lambda^2 P_{tx}}{\alpha Sen}}.$$

$$(7.3)$$

T_S is the sensing range, and the interference range is defined as $T_I = T_R \sqrt[\beta]{TR_{CP}}$, where TR_{CP} and Sen are defined as the threshold of capture ratio and the receiver sensitivity in watts, respectively [36].

f. The propagation model is defined as

$$P_{tx} = Sen \left[\frac{G_t G_t \gamma^2}{(4\pi)^2 d^\beta} \right]$$

$$(7.4)$$

and

$$P_{rx} = \frac{P_{tx}}{\alpha d^\beta},$$

$$(7.5)$$

where d is defined as the distance between two transmission nodes. This model is covered by a free space model which is defined as follows:

$$\alpha = \frac{(4\pi)^2}{\omega^2 G_r G_t},$$

(7.6)

where both G_t and G_r are defined as the antenna gain, γ is the wavelength, and β is defined as the path loss exponent [36].

Table 7.1 lists all assumptions that are used throughout the chapter. A node may exchange its status slot by slot, which corresponds to the transition

Table 7.1 Assumptions

Symbol	Quantity
N	Number of nodes in the network
T	Length of a cycle
W	Contention window size in units of a time slot
M	Queue capacity in units of a data packet
m	Number of data packet in queue
S	Data packet size
i, j	The state of node
$P_{i,j}$	The probability of transition from state i to state j for the node
p_f	Probability of transmission failure of data packet
p_s	Probability of successfully transmission of data packet
n	Number of sensor nodes in specified area
ρ	Density of sensor nodes
T_r	Transmission range
T_s	Sensing range
T_I	Interference range
λ	Expected data packet arrival rate at the MAC layer
A_i	Probability of i data packets arriving in a cycle
π	The stationary distribution of the Markov model
τ	The probability of an arbitrary node transmitting in time slot
D	Packet delay

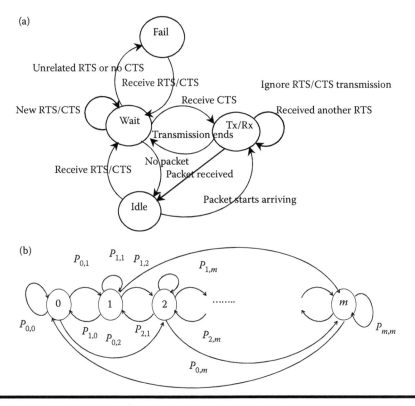

Figure 7.4 Markov model for the X-MAC protocol and node.

from one state to another in the Markov chain as depicted in Figure 7.4a. Figure 7.4b shows that the proposed Markov model has limited queuing capacity (denoted as *M*), with finite state slots from left to right, which corresponds to a 0 state for processing packets in the queue and so on to *m* packets in the queue (full queue). Specifically, if a packet arrives and the queue is full, then the packet is simply dropped; nevertheless, the packets are removed from the queue when they are successfully transmitted. By contrast, when the queue is neither full nor empty, then a node may obtain access to the media to transmit packets with an independent probability. The analysis of the Markov discrete-time M/M/1 queuing model offers insights into the traffic behavior of VSNs in general and points to an idea for a control algorithm. The steady-state probability and the transition probabilities of moving from one state to another can be described as follows:

$$P_{0,i} = \lambda_i \delta, \quad i = 0,\dots,M, \tag{7.7}$$

$$P_{0,i} = A_i, \quad i = 0,\dots,M, \tag{7.8}$$

$$P_{0,M} = A_{\geq M}, \tag{7.9}$$

$$P_{i,i-1} = p_s A_0, \quad i = 0,\dots,M. \tag{7.10}$$

$$P_{i,i-1} = p_s A_{j-i+1} + (1 - p_s) A_{j-i}, \quad i = 1,\dots,M-1. \tag{7.11}$$

$$P_{i,M} = p_s A_{\geq M-i+1} + (1 - p_s) A_{\geq M-i}, \quad i = 1,\dots,M. \tag{7.12}$$

$$P_{i,j} = 0, i = 2,\dots,M, j = 0,\dots,i-2, \tag{7.13}$$

If an abnormal event of interest is detected during the specified operations, then Equations 7.7 and 7.8 describe all transitions from an empty queue status to a non-empty status according to the Poisson process probability of a new packet arrival λ. The typical schedule-driven operation for vehicle-based sensor nodes operates with two timers: one for wake-up mode and another for sleep mode for each node in the network [35]. Therefore, if an abnormal event is detected by a sensor node and needs to be transmitted to another node or to the sink, the node stops the sleep mode timer, turns on its radio, and starts processing the event; otherwise, the node remains in sleep mode. Equations 7.9 and 7.12 describe the transition probability of the schedule-driven duty-cycle node operation, including the processing and transmission of information packets.

Equations 7.10 and 7.11 also describe the non-transition probability state (i.e., the probability of having a non-decreasing queue), which can be obtained from two terms depending on the oldest information packets still in the queue and winning the contention to access the media (first term) or otherwise (second term) [24,36].

7.4.1 The Hidden-Problem Formulation

According to the heavy-traffic assumption [39], each node in the network always has a packet in its buffer to be sent. Suppose a node is ready to transmit with probability p_s, the probability of collision is p_f, A_i defines the probability that i of data packet arrives at the node during a cycle, and $A_{\geq i}$ is the probability that no less than i data packets arrive at a node during a cycle. Then, p_s is considered as a protocol-specific parameter that is slot-independent. This means that the probability of transmission and collision varies from one time slot to another depending on the behavior of both duty-cycled nodes, which are modeled as a Markov chain and the state of the channel, depicted in Figure 7.4, where π_0 indicates the steady state of a node. A node may or may not transmit in the slot depending on the mechanism used to avoid and resolve the collision as well as the current state of the channel [37]. Therefore, the exact relationship between p_s and p_f should be derived to investigate the effects of p_s and p_f on performance of a multi-hop network performance.

Generally, for single-hop IEEE 802.11, Bianchi [28] has proposed a multi-dimensional Markov chain model to derive the probability of successful packet transmission when a node always has a packet in its buffer ready to send and when the next hop is randomly selected as

$$p_s(p_f) = \frac{2}{1 + W + p_f W \dfrac{(2p_f)^m - 1}{2p_f - 1}},$$ (7.14)

where W is the minimum contention window size and m is the retry limit. Supposing that the sensor node is in a steady state, the probability of an arbitrary sensor node transmitting in the time slot denoted by τ is [40]

$$\tau = \frac{p_s T}{D},$$ (7.15)

$$D = \lambda + p_s T_{tran} + Y,$$ (7.16)

where λ defines the average event inter-arrival rate, T_{tran} is the transmission time, and Y defines the processing time per event.

In general, τ depends on the conditional transmission probability p_s and the collision probability p_f, which is still unknown, as shown in Equation 7.3. Because each node n transmits with probability [41] τ,

$$p_s = n\tau(1 - \tau)^{(n-1)}.$$ (7.17)

Moreover, to find the value of p_f in a time slot in which at least one n is still transmitting, this yields [42]

$$p_f = 1 - (1 - \tau)^n - 1.$$ (7.18)

Equations 7.5 and 7.6 represent a nonlinear system in the two unknowns τ and p_s, which are solved by using the Gregory–Newton forward difference approach [43].

Consequently, it can use a similar approach to build the Markov chain model for the duty-cycle MAC protocol of a saturated node to obtain the relationship between p_s and p_f because of low duty-cycle MAC protocols using the same factors as IEEE 802.11 protocols to quantify the QoS parameters. However, there is an extra delay in the low duty-cycle MAC protocol, which is caused by the sleep period of each node [44].

7.4.2 Media Access Rules of X-MAC

The hidden problem considers the first MAC problem that should be solved to investigate the network performance and to provide resource reservation and fulfill QoS requirements. Suppose that the transmission in the sensor networks begins from node A and transmits data packets to node B, as depicted in Figure 7.5. Then, the propagation model between nodes A and B is given as

$$P_{rx}(B) = \frac{P_{tx}(A)}{\alpha d(A,B)^{\beta}}.$$ (7.19)

Therefore, the set of sensor nodes that are able to detect the transmission of node A is denoted as $N_{tx}(A)$ and defined as

$$N_{tx}(A) = \{x \mid d(x,A) \le R\},$$ (7.20)

where R is the transmission range defined as

$$R = \sqrt[\beta]{\frac{P_{tx}(A)}{\alpha T_R}}.$$ (7.21)

Where the sensor nodes are distributed inside the dotted circle, the set of nodes N_{rx} that is distributed outside E transmission range, defined as $E = \sqrt[\beta]{\dfrac{P_{tx}(A)}{\alpha Sen}}$, cannot

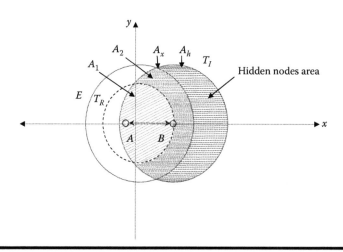

Figure 7.5   Transmission, reception, and interference ranges.

detect the transmission of node A [36]. This set is delimited by the shaded area and is calculated by using the geometry equations as noted in [1].

$$A_H(T_R) = \pi T_I^2 - 2T_I^2 \left[\arccos\left(\frac{T_R}{2T_I}\right) - \frac{T_R}{2T_I}\sqrt{1 - \left(\frac{T_R}{2T_I}\right)^2} \right]. \qquad (7.22)$$

Assume that $A_x(T_R)$ is the common shaded area that illustrates the locations at which possible hidden terminals reside. These circles of radii E and T_I intersect at two points: $(u, -\sqrt{E^2 - u^2})$ and $(u, \sqrt{E^2 - u^2})$, where $u = \dfrac{E^2 + T_R^2 - T_I^2}{2T_R}$. Therefore,

$$A_x(T_R) = [A_1(T_R) + A_2(T_R)], \qquad (7.23)$$

where

$$A_1(T_R) = \int_{-T_I + T_R}^{u} \sqrt{T_I^2 - x^2}\, dx = T_I^2 \left[\frac{\pi - a_2}{2} + \frac{\sin 2a_2}{4} \right] \qquad (7.24)$$

and

$$A_2(T_R) = \int_{u}^{E} \sqrt{E^2 - x^2}\, dx = E^2 \left[\frac{a_3}{2} + \frac{\sin 2a_3}{4} \right], \qquad (7.25)$$

where $a_2 = \arccos\dfrac{u - T_R}{I_R}$ and $a_3 = \arccos\dfrac{u}{E}$. The average value of the hidden nodes are calculated as [45]

$$A_h(T_R) = [\pi I_R^2 - A_x(T_R)]. \qquad (7.26)$$

Hidden sensor nodes distributed inside the shaded area that illustrates the locations at which possible hidden terminals reside may limit the performance of contention-based MAC protocols. Because of their transmission results in collisions, the proposed model quantified the problem by deriving the probability of collision for avoiding this limitation.

The axioms of probability used to estimate the collision probability are performed by evaluating the dimensions of a probabilistic combination of events that might occur within the A_x and A_h areas. Therefore, the derivation of collision probability depends on two rules: the addition rule (which deals with the probability of union of more events) and the multiplication rule (which deals with the probability of intersection of two events).

Suppose that E_x is an event during which collision occurs between one or more nodes within the A_x area, and E_h is an event during which collision occurs between one or more nodes within the A_b area. Then, the collision probability p_f is calculated as [1]

$$p_f = p_{E_x} + p_{E_b} - p_{E_x} p_{E_b}, \qquad (7.27)$$

where p_{E_x} is the probability of the event E_x which is given as [1]

$$
\begin{aligned}
p_{E_x} &= Pr\{\text{two or more two awake sensor nodes in } A_x\} \\
&= \sum_{j=2}^{\infty} \left(\sum_{i=j}^{\infty} \binom{i}{j} p_s^j \, p(i, A_x) \right) \qquad (7.28) \\
&= 1 - (1 + p_s \rho A_x) \exp^{-\rho A_x},
\end{aligned}
$$

where $p(i, A_x) = \dfrac{(\rho A_x)^i}{i!} \exp^{-\rho A_x}$ [1]. Likewise p_{E_b} is defined as the probability of the event E_b that can be obtained as in Reference 1 by

$$
\begin{aligned}
p_{E_b} &= Pr\{\text{only one awake sensor node within } A_x\} \\
&= Pr\{\text{two or more two awake sensor nodes in } A_b\} \\
&= p_s \rho A_x (1 - \exp^{-(1-p_s)\rho A_x}) \qquad (7.29) \\
&\quad (1 - \exp^{-\rho A_b (1-(1-p_s))}) \exp^{-p_s \rho A_x}.
\end{aligned}
$$

7.5 QoS Parameters Analysis of the X-MAC Protocol

QoS parameters expressed in terms of energy consumption, delay, and throughput can be calculated within a cycle time, since X-MAC works in a duty-cycled fashion. Therefore, an active period of a wake-up node is defined in cycle time as T_{awake} time units. During this active period, the preamble packets, data packet with size S, and an ACK message are assumed as having T_{pre}, T_{Data}, T_{ACK} time slot units to transmit, respectively; hence, the Markov model has a unique stationary distribution $\pi = \pi_0, \ldots, \pi_M$ [29].

Usually, X-MAC uses a fixed preamble size carrying the address of the sink to transmit the data packet. Suppose that a sensor network is fully connected with n nodes and then that each node wakes up periodically for successful transmission of a data packet with the probability of $(1 - \pi_0) p_s$, which takes a time on average of $(T/2) + T_{Data}$. Each node uses $(T/2)$ periodically to send preamble packets, and then a node starts to listen to the ACK messages between two successive preamble packets, whereas T_{Data} is used periodically to successfully transmit a data packet with a specific probability. Hence, the average time that it takes to send a preamble packet is $(T/2)(T_{pre}/(T_{pre}+T_{ACK}))$. The average time a node takes to listen to the media is $(T/2)(T_{ACK}/(T_{pre}+T_{ACK}))$. Finally, the amount of energy that is consumed in this case is calculated as [18]

$$\text{Energy}_1 = (1-\pi_0)p_s\tau\left(\frac{T}{2}\left(\frac{T_{pre}}{T_{pre}+T_{ACK}}\right)\right)P_{TX}$$

$$+\frac{T}{2}\left(\frac{T_{ACK}}{T_{pre}+T_{ACK}}\right)P_{RX}+T_{Data}P_{RX}. \qquad (7.30)$$

In a similar case, a node wakes up periodically to an unsuccessful transmission of a data packet with probability $(1-\pi_0)p_f$. However, in this case, there is no preamble packet that may be received correctly because of a collision, and then the node continues sending preamble packets with an average time of $T(T_{pre}/(T_{pre}+T_{ACK}))$ and an average time of listening to media between two successive preambles of $T(T_{ACK}/(T_{pre}+T_{ACK}))$. The amount of energy consumed in this case is [18]

$$\text{Energy}_2 = (1-\pi_0)p_f\tau\left\{\left[T\left(\frac{T_{pre}+T_{ACK}}{2P_{RX}+T_{pre}P_{RX}+T_{ACK}P_{TX}+T_{DATA}P_{RX}}\right)\right]\right\}. \qquad (7.31)$$

Suppose that the node has received complete preamble packets. It then sends back a T_{ACK} message to receive the data packet. However, any intermediate nodes may wake up to send a preamble packet or listen for receiving T_{ACK} messages from the sink. As a result, the time on average of the receiving node is $((T_{pre}+T_{ACK})/2)$. Therefore, the amount of energy consumed is

$$\text{Energy}_3 = (1-\pi_0)p_f\tau\left(T\left(\frac{T_{pre}}{T_{pre}+T_{ACK}}\right)\right)$$

$$P_{TX}+T\left(\frac{T_{ACK}}{T_{pre}+T_{ACK}}\right)P_{RX}. \qquad (7.32)$$

In case a node is in a wake-up state while the colliding preamble packets are half-way to transmission or when the nodes are listening to the media, the node does not go to sleep and stays awake for the time average $((T_{pre}+T_{ACK})/(2+T_{pre}))$. Because the node cannot detect the collision until it hears the next colliding preamble packets. Hence, the amount of energy consumed is

$$\text{Energy}_4 = (1-\pi_0)p_f\tau\left(\frac{(T_{pre}+T_{ACK})}{2P_{RX}+T_{pre}P_{RX}}\right). \qquad (7.33)$$

Finally, the energy consumption per hop per second can be obtained as

$$\text{Energy} = \sum_{i=1}^{4} E_i. \tag{7.34}$$

The analysis of the delay of a data packet depends on the probability of successful transmission p_s for each node that wins the contention and the steady state of the Markov model π_0 during the cycle. Therefore, a node starts to send contentions for the media during the cycle until it wins the contention with probability p_s or loses the contention with probability $1 - p_s$. Hence, the delay is given along cycle length T:

$$\text{Delay} = T \sum_{i=1}^{\infty} (i+1) p_s (1 - p_s)^i. \tag{7.35}$$

The derivation of the probability of successful transmission and collision in the X-MAC protocol depends on the status of the queue and the channel. This means that a node has an *empty/empty* queue according to the steady state *free/busy* of the proposed Markov chain model. Further, the status of the channel is *free/busy* according to the detection of the collision occurring by event. Suppose that A defines an event that occurs by one or more node winners in contention within the area A_x and B is defined as an event that occurs between two or more node winners in contention with area A_b. It follows that [9]

$$p_s = P_r(A, free|empty) = P_r(A| free, \overline{empty}) P_r(free|\overline{empty}) \tag{7.36}$$

and

$$p_f = P_r(B, free|empty) = P_r(B| free, \overline{empty}) P_r(free|\overline{empty}). \tag{7.37}$$

To solve for $P_r(A, free|empty)$ and $P_r(B, free|empty)$, assume that a node has an empty queue, directly wakes up, detects the event, and transmits a data packet for its neighbors. At the same time, other nodes are waking up but do not have any data packets for transmission. Therefore,

$$P_r(A| free, \overline{empty}) = \sum_{t=1}^{T} \frac{1}{T} \left[\sum_{i=1}^{N_c-1} \binom{N_c-1}{i} \frac{1}{T}^i \pi_0^i \left(\frac{T-1}{T} \right)^{N_c-1-i} \right]. \tag{7.38}$$

Similarly, when one other node wakes up and has data packet for transmission, then a collision occurs; therefore, the collision probability is

$$P_r(B| free, \overline{empty}) = \sum_{t=1}^{T} \frac{1}{T} \left[\sum_{i=1}^{N_c-1} \binom{N_c-1}{i} \left(\frac{1}{T} \right)^i \left[\sum_{j=1}^{i} (1-\pi_0)^j \pi_0^{i-j} \left(\frac{T-1}{T} \right)^{N_c-1-i} \right] \right].$$

$$\tag{7.39}$$

$P_r(free|empty)$ is defined as the probability of a free channel when a node wakes up and has a data packet in its queue ready for transmission. As mentioned, each node in the X-MAC protocol periodically wakes up and sends preamble packets, and then a node starts to listen to the channel. The channel has the same probability of being free or busy in every time slot [18]. Hence, when a node wakes up, no matter whether its queue is empty or not, the node sees the channel with the same probability of being free or busy. Therefore,

$$P_r(free) \approx P_r(free|empty). \tag{7.40}$$

The probability of a free channel is determined by two parameters: (a) the average length of the free channel and (b) the average length of the busy channel. Therefore, the probability of a free channel is expressed as

$$P_r(free) = \frac{C_{free}}{C_{free} + C_{busy}}. \tag{7.41}$$

To calculate C_{free}, assume that the interval of time transmission ends and the free channel begins. There are a number of cycles $cycle$ that could define the free channel until some nodes begin to transmit at t^{th} slot. Thus, the average length of a free channel is $cycleT + t$, and the probability of the transmission event is [18]

$$P_{free}(cycle,t) = \pi^{N_c cycle} \sum_{i=0}^{N_c-1}$$

$$\sum_{j=1}^{N_c-j} \sum_{k=1}^{j} \binom{N_c}{i} \left(\frac{t}{T}\right)^i \pi_0^i$$

$$\binom{N_c-i}{j} \left(\frac{1}{T}\right)^j \binom{j}{k} (1-\pi_0)^k \pi^{j-k} \left(\frac{T-t-1}{T}\right)^{N_c-i-j}. \tag{7.42}$$

Therefore, the average length of a free channel can be obtained as [18]

$$C_{free} = \sum_{cycle=0}^{\infty} \sum_{t=0}^{T-1} (cycleT + t) * P_{free}(cycle,t). \tag{7.43}$$

The successful transmission probability P_{suc} could be similarly obtained as the probability of a collision P_{col}, as long as the channel is free for $cycle$ cycles and t slots. Therefore, both successful transmission and collision are calculated as [18]

$$P_{suc}(cycle,t) = \pi_0^{N_c} \sum_{i=0}^{N_c-1} \sum_{j=1}^{N_c-i} \binom{N_c}{i} \left(\frac{t}{T}\right)^i$$

$$\pi_0^i \binom{N_c-i}{j} \left(\frac{1}{T}\right)^j \binom{j}{1} (1-\pi_0) \pi^{j-1} \left(\frac{T-t-1}{T}\right)^{N_c-i-i}. \tag{7.44}$$

$$P_{col}(cycle,t) = \pi_0^{N_c} \Sigma_{i=0}^{N_c-2} \Sigma_{j=2}^{N_c-i} \Sigma_{k=2}^{j} \binom{N_c}{i} \left(\frac{t}{T}\right)^i$$

$$\pi_0^i \binom{N_c-i}{j} \binom{j}{k}^j (1-\pi_0)^k \pi_0^{j-k} \left(\frac{T-t-1}{T}\right)^{N_c-i-j} \quad (7.45)$$

The average length of a busy channel can be calculated according to the successful transmission of a data packet, which takes on average time $(T/2) + T_{Data} + P_{suc}$. Therefore,

$$C_{busy} = \sum_{cycle=0}^{\infty} \sum_{t=0}^{T-1} \left[\left(\frac{T}{2} + T_{DATA} \right) P_{suc}(cycle,t) + TP_{busy}(cycle,t) \right]. \quad (7.46)$$

By plugging Equations 7.44 and 7.41 into 7.39, $P_r(free)$ is obtained. By plugging 7.36 and 7.38 into 7.34 and then plugging 7.37 and 7.38 into 7.35, the successful transmission probability for each node p_s and the probability of collision p_f are obtained. Therefore, the throughput per-second per-hop of the X-MAC protocol can be calculated as

$$\text{Throughput} = N_c(1-\pi_0)p_s \frac{S}{(T\tau)}. \quad (7.47)$$

7.6 Simulation Results

Consider a vehicle equipped with wireless devices; therefore, they can communicate with each other. However, the primary application is to let vehicles exchange their current context in order to detect abnormal events in an urban environment. The network topology is shown in Figure 7.6, in which our proposed model provides detailed analysis for the performance of a synchronized X-MAC protocol under the impact of the hidden problem as well as under unicast traffic for various network configurations, conditions, and other assumptions for specific scenarios. The analytical correctness of the proposed model was validated through implementation by using MATLAB [46].

In all the simulations to be presented in the chapter, the network setup is fully connected with a varying number of connected nodes (that is, vehicles). The transmission range of each vehicle is 18 m, and the cycle length of each time period is 50 s. Simulation, energy consumption, delay, and throughput are investigated for each set under varying network parameters of the vehicle. The values of the network parameters used are summarized in Table 7.2. These parameters are set to comply with the X-MAC protocol specifications. All vehicles are distributed by rating by using a two-dimensional Poisson distribution within an area of 50 * 50 m².

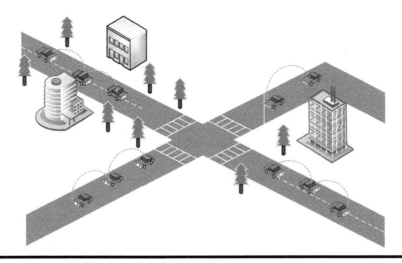

Figure 7.6   Network topology to evaluate the X-MAC protocol.

Figure 7.7 shows the performance of an X-MAC protocol for the multi-hop communication fashion under varying cycle length T from 50 to 200 msec, as shown in Figure 7.8. Figure 7.9 shows the performance of an X-MAC protocol under varying vehicle density in the network and transmission range of each vehicle varying from 18 to 29 m, as shown in Figure 7.10. From these results, it can be seen that the analytical results of the proposed model match the simulation results.

7.6.1 Varying the Cycle Length

In this experiment, we vary the cycle length. Figure 7.7a shows the power consumption of the X-MAC protocol obtained from simulations and from our power consumption analysis using the Markov model. Our analytical results match the simulation results with percentage difference or maximum difference in power consumption. The power consumption decreases as the length of the cycle period increases, which means that whenever the cycle length increases, the unsuccessful data transmission increases due to an increase in the rate of collision probability. Consequently, as the cycle length increases and as the number of vehicles increases, the power consumed in data transmission increases, and the active period in each cycle is fixed. Therefore, all vehicles in the network expect the source vehicle and the sink node in either successful transmission or unsuccessful transmission to go to the sleep state as long as the cycle length increases. Thus, the power saving in this longer sleep period is greater than that in the longer transmission period; the power consumption of the X-MAC protocol decreases as the cycle length increases.

Figure 7.7b shows that the average delay of the X-MAC protocol increases as the length of the cycle period increases, usually before the X-MAC protocol

Table 7.2 Experiment Parameters for the X-MAC Protocol

Parameter	Value
T	200 msec
t_{data}	50 msec
t_{sync}	30 msec
t_{RTS}	15 msec
t_{CTS}	15 msec
t_{ACK}	5 msec
Topology structure	Square (50 * 50 m²), sensor node distributed uniformly
Total number of sensor nodes	5, 12, 17 sensor nodes
Message payload	64 bytes
Node density	0.002, 0.005, 0.007
Tx data rate	250 kbps
Transmission range	18–29 meter
Propagation delay	1 μsec
tx_{xp}	0.0525 watts
rx_{xp}	0.0591 watts
sp	0.0525 watts

saturates, and the cycle length starts from a small value with a few vehicles in the network. The X-MAC protocol can deliver all the incoming data packets as soon as they arrive in the network; hence, the delay is nearly zero. Once the X-MAC protocol saturates and the cycle length increases, both the contention and queue delay increase proportionally with the increases of the cycle length and the number of nodes. Since the X-MAC protocol can no longer deliver all the incoming data packets, the queue at each vehicle overflows and data packets are dropped.

Figure 7.7c reveals that the average throughput of the X-MAC protocol decreases as the cycle length increases and as the number of vehicles in the network increases. This means that before X-MAC saturates, the number of incoming data packets that are delivered remains in the queue and slightly decreases when the X-MAC protocol saturates and as the number of vehicles in network increases further. The X-MAC protocol can no longer deliver all the incoming data packets because of the increasing number of collisions at the following specified period.

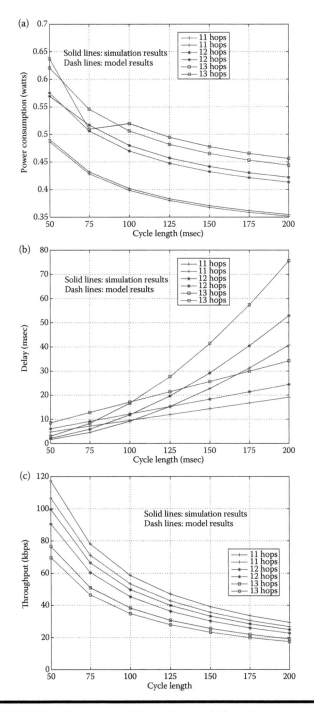

Figure 7.7 X-MAC performance with varying cycle length.

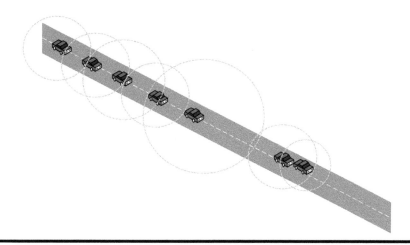

Figure 7.8   First experiment to evaluate the X-MAC protocol.

7.6.2 Varying the Number of Nodes

In this experiment, we vary the number of vehicles. Figure 7.9 shows the power consumption of the X-MAC protocol increases under varying vehicle density in the network and transmission range of each vehicle in the network as seen in Figure 7.10. Whenever the transmission range increases, the unsuccessful data transmission increases, since the rate of collision probability increases and more power is consumed. Consequently, as the transmission range increases and as the number of vehicles increases, the power consumed in data transmission increases because the active period in each cycle is fixed. Therefore, all vehicles in the network expect the main vehicle source and the main sink node in either successful transmission or unsuccessful transmission to wake up as long as the cycle length increases. Thus, power savings in this longer sleep period become greater than in the longer transmission period. The power consumption of the X-MAC protocol decreases as the cycle length increases.

Figure 7.9b shows the delay of the X-MAC protocol. Before X-MAC saturates, the delay slowly increases with the lower node density and with fewer variations in the transmission range vehicle in the cycle. Moreover, as the number of vehicles increases, a vehicle tends to have a higher probability of successful transmission leading to the saturation state. Therefore, the delay dramatically increases as X-MAC saturates because of the increases in the rate of collision probability in a cycle. This means that more vehicles in the network leads to decreasing the probability of successful transmission of packets when X-MAC saturates. Hence, according to the main parts of the X-MAC protocol data-packet delay (the queue delay and contention delay), this increase is obvious when the number of vehicles and the transmission range increase.

Figure 7.9c shows the throughput of the X-MAC protocol. Before X-MAC saturates, the number of incoming data packets in the cycle remains, and the node can

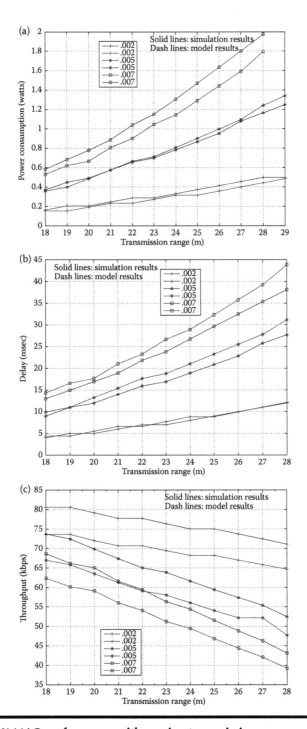

Figure 7.9   X-MAC performance with varying transmission range.

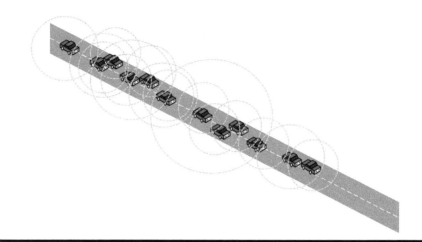

Figure 7.10 Second experiment to evaluate the X-MAC protocol.

no longer deliver all incoming data packets in the cycle. Once the transmission range of a vehicle in a cycle increases, the throughput decreases and the vehicle density increases. This means that X-MAC can no longer deliver all incoming data packets because of an increase in the transmission range. A vehicle tends to have a higher probability of successful packet transmission, which leads to a saturation state. Therefore, the throughput shrinks dramatically because of the increase in the probability of collision.

7.6.3 Discussion

The proposed Markov model for asynchronous duty-cycled MAC protocols has been applied in the X-MAC protocol to optimize some protocol parameters to achieve desirable performance. Conceptually, X-MAC is considered a more efficient protocol than the synchronized S-MAC protocol because of the avoidance of synchronization overheads. Therefore, asynchronous X-MAC can achieve more desirable performance than synchronized S-MAC for the delay and throughput [47]. But this does not mean that asynchronous X-MAC does not suffer from hidden terminals, especially when varying the vehicles' density. As the density of the vehicles increases, more than one vehicle wakes up and simultaneously start to send their preambles. Therefore, other vehicles cannot determine whether the information is in preambles and so collisions occur. Consequently, and based on the assumption used in the Markov model in the synchronous MAC protocol, our proposed model uses two different values of probabilities for transmission and collision (p_s and p_f, respectively) to handle the hidden terminal problem and to estimate the communication links' performance. Finally, we should distinguish between the effect of asymmetric and symmetric links in order to provide the adequate functionality to routing protocols. Asymmetric links are caused by several factors, like node mobility, heterogeneous radio technologies, and irregularities in radio ranges. Although it is difficult to achieve high network connectivity, high data rate transmission, and low latencies, according to our observations, it is expected that asymmetric links will be more common in the future than VSNs. Actually, the hidden terminal problem will be more complicated with the existence of asymmetric links, where the receiver and sender do not share the channel feedback and hidden nodes may interfere with the ongoing transmissions [48]. Furthermore, because of the feedback from the receiver in RTS/CTS, exchanged messages may have to pass through several relay nodes before being delivered to all others nodes. However, MAC protocols might need to exploit asymmetric links to solve the hidden terminal problem while maintaining the lowest cost. In fact, network utilization and cost are weights derived from and constrained by the network resource management (NRM) approach, in which some routing or MAC protocols based on QoS parameters (such as power consumption, throughput, and latency) are application-dependent and imposed by the node specification. Therefore, to optimize these parameters in order to achieve the desirable benefits and network performance, the NRM approach should be chosen based on the desired QoS parameters. As a result, the inclusion of asymmetric links in the MAC protocol design can further improve the network performance.

7.7 Conclusion

In this chapter, the QoS parameters of X-MAC are modeled and analyzed by using the Markov chain model. Our model presents analytical results that have been validated by simulation results for the selected QoS parameters under various network conditions and traffic loads. It provides sufficient information about the links between the vehicles to determine the optimal path and to select the intermediate nodes for packet routing from source to distention. Our future work will focus on analyzing the QoS parameters for synchronized protocols and extending the model to multi-hop networks.

7.8 Competing Interests

The authors declare no conflict of interest.

7.9 Funding

The authors declare that they depend on their own personal funding in order to publish this article.

7.10 Author's Contributions

Mohammed Zaki Hasan and Fadi Al-Turjman conceived the idea and wrote the chapter. Mohammed Zaki Hasan performed the experiments and analyzed the data. Fadi Al-Turjman gave valuable suggestions on the motivation of this work and assisted in revising and proofreading the chapter.

Acknowledgments

The authors extend their appreciation to the anonymous reviewers for their helpful and supportive comments toward improving our chapter.

References

1. F. Al-Turjman, *Cognitive-Node Architecture and a Deployment Strategy for the Future Sensor Networks*, Springer Mobile Networks and Applications, 2017, doi: 10.1007/s11036-017-0891-0
2. F.M. Al-Turjman, Information-centric sensor networks for cognitive IoT: An overview, *Annals of Telecommunications*, pp. 1–16, 2016.
3. G.T. Singh and F.M. Al-Turjman, Learning data delivery paths in QoI-aware information-centric sensor networks, *IEEE Internet of Things Journal*, vol. 3, no. 4, pp. 572–580, 2016.
4. F. Cunha et al., Data communication in VANETs, *Ad Hoc Networks*, vol. 44, pp. 90–103, 2016.
5. K. Nellore and G.P. Hancke, A survey on urban traffic management system using wireless sensor networks, *Sensors*, vol. 16, no. 2, p. 157, 2016.
6. F. Al-Turjman and M. Gunay, CAR approach for the internet of things, *Canadian Journal of Electrical and Computer Engineering*, vol. 39, no. 1, pp. 11–18, 2016.
7. W. Xiang et al., Wireless access in vehicular environments, *EURASIP Journal on Wireless Communications and Networking*, vol. 2009, no. 1, pp. 1–2, 2009.
8. G.P. Hancke, and G.P. Hancke Jr., The role of advanced sensing in smart cities, *Sensors*, vol. 13, no. 1, pp. 393–425, 2012.
9. T. Asha and N. Muniraj, Energy efficient topology control approach for mobile ad hoc network, *International Journal of Computer Science Issues (IJCSI)*, vol. 10, no. 4, 289, 2013.
10. U. Lee and M. Gerla, A survey of urban vehicular sensing platforms, *Computer Networks*, vol. 54, no. 4, pp. 527–544, 2010.
11. Lee, U. et al., Dissemination and harvesting of urban data using vehicular sensing platforms, *IEEE Transactions on Vehicular Technology*, vol. 58, no. 2, pp. 882–901, 2009.
12. P. Suriyachai, U. Roedig, and A. Scott, A Survey of MAC protocols for mission-critical applications in wireless sensor networks. *Communications Surveys & Tutorials, IEEE*, vol. 14, pp. 240–264, 2012.
13. N. Saxena, A. Roy, and J. Shin, Dynamic duty cycle and adaptive contention window based QoS-MAC protocol for wireless multimedia sensor networks. *Computer Networks*, vol. 52, pp. 2532–2542, 2008.
14. P. Mohapatra, *Ad Hoc Networks: Technologies and Protocols*. Springer Science & Business Media, New York, USA, 2005.

15. M. Z. Hasan, H. Al-Rizzo, and F. Al-Turjman, A Survey on multipath routing protocols for QoS assurances in real-time multimedia wireless sensor networks, *IEEE Communications Surveys & Tutorials*, 2017, doi: 10.1109/COMST.2017.2661201

16. P. Park et al., Duty-cycle optimization for IEEE 802.15.4 wireless sensor networks, *ACM Transactions on Sensor Networks (TOSN)*, vol. 10, no. 1, p. 12, 2013.

17. M. Buettner, G. V. Yee, E. Anderson, and R. Han, X-MAC: A short preamble MAC protocol for duty-cycled wireless sensor networks, In *Proceedings of the 4th International Conference on Embedded Networked Sensor Systems*, Boulder, CO, USA, 2006, pp. 307–320.

18. Y. Ou and W. Heinzelman, Modeling and performance analysis for duty-cycled MAC protocols with applications to S-MAC and X-MAC, *IEEE Transactions on Mobile Computing*, vol. 11, no. 6, pp. 905–921, 2012.

19. A. Iyer and C. Rosenberg, Understanding the key performance issues with MAC protocols for multihop wireless networks. *Wireless Communications and Mobile Computing*, vol. 6, pp. 745–760, 2006.

20. G. Wang et al., A MAC layer protocol for wireless networks with asymmetric links, *Ad Hoc Networks*, vol. 6, no. 3, pp. 424–440, 2008.

21. Y. Wei, J. Heidemann, and D. Estrin, An energy-efficient MAC protocol for wireless sensor networks, in *The 21st Annual Joint Conference of the IEEE Computer and Communications Societies*, New York, USA, June 23–27, 2002.

22. T. Dam and K. Langendoen, An adaptive energy-efficient MAC protocol for wireless sensor networks, in *Proceedings of the First International Conference on Embedded Networked Sensor Systems*, Los Angeles, California, USA, pp. 171–180, 2003.

23. J. Polastre, J. Hill, and D. Culler, Versatile low power media access for wireless sensor networks, in *Proceedings of SenSys'04*, Baltimore, MD, USA, pp. 95–107, 2004.

24. M. R. Ahmad, E. Dutkiewicz, and X. Huang, A survey of low duty cycle MAC protocols in wireless sensor networks, in A. Foerster and A. Foerster (eds.) *Emerging Communications for Wireless Sensor Networks*, InTech: Raleigh, NC, pp. 69–90, 2011, ISBN 978-953-307-082-7.

25. A. Razaque and K. M. Elleithy, Energy-efficient boarder node medium access control protocol for wireless sensor networks. *Sensors*, vol. 14, pp. 5074–5117, 2014.

26. L. Alazzawi and A. Elkateeb, Performance evaluation of the WSN routing protocols scalability, *Journal of Computer Systems, Networks, and Communications*, vol. 2008, Article ID 481046, 9 pages, 2008. doi:10.1155/2008/481046

27. H. Wu, R. M. Fujimoto, and G. Riley, Experiences parallelizing a commercial network simulator, in *Proceedings of the Winter in Simulation Conference*, Arlington, VA, USA, pp. 1353–1360, 2001.

28. G. Bianchi, Performance analysis of the IEEE 802.11 distributed coordination function. *IEEE Journal on Selected Areas in Communications*, vol. 18, pp. 535–547, 2000.

29. Y. Ou and W. B. Heinzelman, Modeling and throughput analysis for X-MAC with a finite queue capacity, in *2010 IEEE Global Telecommunications Conference (GLOBECOM 2010)*, Miami, FL, USA, pp. 1–5, 2010.

30. I. Rhee, A. Warrier, M. Aia, J. Min, and M. L. Sichitiu, Z-MAC: A hybrid MAC for wireless sensor networks. *IEEE/ACM Transactions on Networking (TON)*, vol. 16, pp. 511–524, 2008.

31. M.Z. Hasan, F. Al-Turjman, and H. Al-Rizzo, Evaluation of a duty-cycled protocol for TDMA-based Wireless Sensor Networks. *International Wireless Communications and Mobile Computing Conference (IWCMC)*, Paphos, Cyprus, pp. 964–969, 2016.

32. G. Yan and D.B. Rawat, Vehicle-to-vehicle connectivity analysis for vehicular ad-hoc networks. *Ad Hoc Networks*, vol. 58, pp. 25–35, 2017.

33. F. Al-Turjman, Hybrid approach for mobile couriers election in smart-cities. *2016 IEEE 41st Conference on Local Computer Networks (LCN)*, Nov. 7–10, Dubai, United Arab Emirates, pp. 507–510, 2016.

34. O. Yang and W. Heinzelman, Modeling and throughput analysis for SMAC with a finite queue capacity, *5th International Conference on Intelligent Sensors, Sensor Networks and Information Processing (ISSNIP)*, Melbourne, VIC, Australia, pp. 409–414, 2009.

35. M.A. Yigitel, O.D. Incel, and C. Ersoy, QoS-aware MAC protocols for wireless sensor networks: A survey, *Computer Networks*, vol. 55, pp. 1982–2004, 2011.

36. A. Bachir, D. Barthel, M. Heusse, and A. Duda, Hidden nodes avoidance in wireless sensor networks, *International Conference on in Wireless Networks, Communications and Mobile Computing*, Maui, HI, USA, pp. 612–617, 2005.

37. J.J. Garcia-Luna-Aceves and Y. Wang. Collision avoidance protocols in Ad Hoc Networks. In P. Mohapatra and S.V. Krishnamurthy (eds.) *Ad Hoc Networks*. Springer, Boston, MA, 2005.

38. Q. Cao, T. Yan, J. Stankovic, and T. Abdelzaher. Analysis of target detection performance for wireless sensor networks. In V.K. Prasanna, S.S. Iyengar, P.G. Spirakis, M. Welsh (eds.) *Distributed Computing in Sensor Systems. DCOSS 2005. Lecture Notes in Computer Science*, vol. 3560. Springer, Berlin, Heidelberg, 2005.

39. A. Redondi et al., Compress-then-analyze vs analyze-then-compress: What is best in visual sensor networks? *IEEE Transactions on Mobile Computing*, vol. 99, pp. 1–1, 2016.

40. F. Al-Turjman, H. Hassanein, and M. Ibnkahla, Towards prolonged lifetime for deployed WSNs in outdoor environment monitoring, *Elsevier Ad Hoc Networks Journal*, vol. 24, no. A, pp. 172–185, Jan. 2015.

41. Y. Kim, C. Yang, and C.-H. Liu, Throughput analysis of randomized sleep scheduling with constrained connectivity in wireless sensor networks, in *Proceedings of the IEEE Global Telecommunications Conference*, New Orleans, LA, USA, pp. 1–6, 2008.

42. D. Rodenas-Herraiz, A.-J. Garcia-Sanchez, F. Garcia-Sanchez, and J. Garcia-Haro, On the improvement of wireless mesh sensor network performance under hidden terminal problems, *Future Generation Computer Systems*, vol. 45, pp. 95–113, 2015.

43. L.M. Milne-Thomson, *The Calculus of Finite Differences*, AMS Bookstore, Providence, RI, 2000.

44. Y. Li, W. Ye, and J. Heidemann, Energy and latency control in low duty cycle MAC protocols, *IEEE Conference in Wireless Communications and Networking*, New Orleans, LA, USA, pp. 676–682, 2005.

45. M. Fewell, *Area of Common Overlap of Three Circles*, Australian Government, Department of Defence, Defence Science and Technology Organisation, Edinburgh, South Australia, October 2006.

46. MathWorks, *M.U.s.G.*, MathWorks. Inc.: Natick, MA, 1992.

47. E. Ziouva and T. Antonakopoulos, CSMA/CA performance under high traffic conditions: Throughput and delay analysis, *Computer Communications*, vol. 25, pp. 313–321, Feb. 2002.

48. G. Wang et al., A simulation study of a MAC layer protocol for wireless networks with asymmetric links, in *Proceedings of the international conference on Wireless communications and mobile computing. ACM*, Vancouver, British Columbia, Canada. pp. 929–936, 2006.

Chapter 8

Mobile Traffic Modeling for Wireless Multimedia Sensor Networks in IoT

8.1 Introduction

Wireless sensor networks (WSNs) are one of the key enabling technologies for Internet of things (IoT) applications that help in gathering and providing the requested information from the environment to the user [1,2]. IoT applications aim mainly at utilizing the most advanced communication technologies in order to provide smart services for citizens in several fields, such as health care, public safety, and transportation. In order to build such a smart paradigm, a huge number of IoT devices (sensors) should be deployed to gather and deliver the massive amounts of data through the network. These devices are equipped with storage, communication, computing, and sensing (imaging) capabilities to enable the network elements to communicate multimedia messages with each other and with the end users. Multimedia sensing is facing many challenges, such as energy consumption, cost, and the inability of the current wireless network infrastructure to deal with its huge amounts of data traffic.

For the first challenge, regardless of how much we increase the number of deployed sensors, they consume even more energy to work, and this can increase the cost of energy. In addition, the reality is that most of the sensors are working with batteries, and more energy consumption can make sensor nodes run out of energy quickly, causing these nodes to die; as a result, the connectivity of the network may be badly affected. For the second challenge, deploying huge numbers of sensors will increase the hardware cost of the network, and the network will not be financially affordable to be applied. Thirdly, in dense sensing, data will be gathered in huge amounts, and current wireless network infrastructures, such as 3G or 4G, can't contain these amounts of data. For instance, the authors in Reference 3 propose an emergent and pedestrian tracking system in which they use opportunistic ad hoc communications to track and evacuate pedestrians in case of a disaster. In this model, pedestrians' smart phones store and carry messages to a number of mobile sinks, which are responsible for communicating with smart phones and getting to the emergent location effectively.

However, IoT applications convey new design challenges for WSNs. Energy consumption, storage management, heterogeneity of devices, and communication bandwidth are some of the major challenges facing this emerging IoT paradigm [4]. So far, WSNs have only been designed to serve application-specific needs. But in the IoT paradigm, they need to provide an application platform with heterogeneous sensors that multiple users can access to gather information about different phenomena, at different instants of time and from different parts of the network. The heterogeneity in sensing devices and user requests generate heterogeneous multimedia traffic flows in the network, which is very difficult for the energy-constrained sensor network to manage. The operation of inefficient design of the antenna is one of those design issues that significantly exhausts most of the battery life of the sensor node. Consequently, it might adversely affect the wireless channel and lead to error-prone links and inefficient routing [5]. The presence of low power-receiving states as well as application-specific protocols such as LoRa and ZigBee can potentially limit battery-draining activity and improve network performance.

Meanwhile, traffic access to the media from the distributed sensor nodes should be strictly controlled to avoid redundancy and collisions, which have a dramatic impact on the lifetime of the wireless multimedia sensor networks (WMSNs) [5]. WMSNs can be used to aggregate multimedia traffic loads and relay to the backhaul Internet in the IoT or other access networks. Unless it is planned carefully, this can result in extensive amounts of energy consumption. Considerable amounts of energy can be wasted unless an appropriate traffic modeling approach is considered for the WMSN nodes that are deployed to serve static/mobile users in IoT environments. Therefore, an accurate traffic modeling strategy is required to predict the performance of the system, especially in terms of energy consumption. This would not be achieved without a realistic case study analysis and an accurate analytical model that can predict the system performance under such setups.

In this chapter, a systematic model is proposed for an upcoming multimedia routing technique, which finds the optimal quality of service (QoS)-based path. This model examines the effects of multi-hop communication (MHC) while deriving a closed form for the bit error rate (BER). To manage the performance of energy consumption and QoS factors, the WMSN topology should match the actual surrounding setups. Consequently, an unwarranted path loss model for adaptive switching between two types of transmission structures is designed. An effective path loss model is considered based on a Markov discrete-time M/M/1 model and applied to the WMSN duty-cycled nodes. The surrounding environment-specific metric in the path loss model, called the degree of irregularity (DOI), has been considered. It is a function of the distance between two nodes. The adaptive switching mentioned above occurs with respect to the DOI. The DOI is used to determine which path will be used to send a packet from the source node to the next hop. The analysis of the energy consumption, delay, and throughput can be then used to optimize data routing protocols in WMSNs. Thus, the main contribution of this chapter is the development of a framework for analyzing the optimal forwarding choices with respect to QoS parameters, while quantifying the impact of the relay of radio irregularity at the medium access control (MAC) layer in two transmission schemes.

The remainder of this chapter is organized as follows. In Section 8.2, an overview of previous analyses of MHC schemes in WMSNs is provided. System models and a Markov queuing model for multimedia routing are introduced in Section 8.3. In Section 8.4, a practical use case is provided, and simulation results for the discussed models are presented. In Section 8.5, the use case performance is evaluated and assessed. Finally, the concluding remarks are summarized in Section 8.4.

8.2 Related Work

Recently, there have been several IoT projects, such as smart cities, smart health care, and smart streets' lighting that are seeking ways to integrate three functions (sensing, processing, and communicating) into a single integrated circuit for various real-time applications with limited energy consumption [6]. Moreover, research about how all these critical processes are going to function together is underway. The author in Reference 7 analyzes the fundamental issue of component interaction and operation under the IoT umbrella. Furthermore, studies focusing on new techniques (such as cooperative multilayer communication among nodes and network coding for wireless communication using particle-sized sensor nodes that are distributed for wide-area sensing) have been proposed in Reference 8. However, the increased interest in real-time applications in multimedia sensor networks have led these studies to focus on enhancing the network performance by relying on an accurate link estimation in order to assure efficient use of energy resources in the sensor node. Cross-layer awareness is considered a potential solution to various issues and a way to improve the performance in WMSNs because of the possibility of involving both the physical (PHY) and MAC layers to provide functions other than routing, such as power efficiency. On this note, we can decide to use asynchronous medium access control (X-MAC) duty-cycled protocols rather than synchronized because X-MACs have lower packet latency and higher energy efficiency due to their reduced idle listening [9].

Recently, many research attempts have been carried out by many researchers on the modeling of WMSN traffic to investigate the performance of the network. These attempts can be classified into the following classes: static, dynamic, and hybrid. In the static modeling approach, WMSN nodes are installed in indoor environments (such as inside a home, office building, shopping mall, or airport) to improve indoor connectivity and enhance the users' Internet experience [10–12]. In the dynamic modeling approach, unlike the static one, sensor nodes are not static and are deployed in moving vehicles to provide services to mobile users. For example, sensor nodes can be installed in public transportation vehicles, such as busses and trains, to enhance coverage and Internet connectivity while on the move [5–7,13]. A number of cooperative/hybrid methods have also been used in WMSNs to model the network performance. These methods are mostly classified as either cooperative automatic repeat query (ARQ) protocols or conversion from single-hop to multi-hop transmission protocols. The authors in Reference 14 propose a network coding-based cooperative ARQ (NCCARQ) MAC procedure for WSNs. This protocol focuses on a centralized WMSN system that manages the retransmission of channel access that provides bidirectional connection between sensor nodes. Every sensor node stores a copy of the packet that has been received until they receive a positive acknowledgment from the sink; otherwise, the error mechanism completes an error check on the received message. Therefore, a channel-sense, multi-access-based protocol forms a good match with the IEEE 802.15.4 standard. Additionally, this compatibility enables the NCCARQ to use the same style to control the packets and follow the same standards, but with added adjustments to enhance the efficiency of the proposed protocol. It is similar to the model proposed in this chapter; it uses fewer control packets than the ARQ-based protocol, with efficient energy use and the fulfillment of QoS parameters.

A typical ad hoc network operates according to a cooperative multi-hop transmission approach, which achieves greater power efficiency because it operates at a low signal-to-noise ratio (SNR) that is needed to cover the transmission range. In Reference 15, authors adopted the linear multi-hop transmission approach by considering quasi-static fading without spatial reuse. This adaptation simplified the linear multi-hop transmission approach by including hybrid automatic repeat request (HARQ) retransmission protocols. The authors focused on a design that provides the optimal number of hops with a maximum delay along a linear multi-hop network that achieves maximum end-to-end throughput. This analytical framework allows the parameters to be set as an optimization problem, which is solved by using numerical methods. Likewise, the authors in Reference 10 considered an uncooperative linear approach for multi-hop and single-hop network transmissions (in which the nodes do not cooperate and instead attempt to access the channel simultaneously) to investigate the performance of the distributed channel access capacity at the MAC layer and the power channel at the PHY layer, especially in delay and bandwidth-constrained scenarios. The analytical framework provides the optimum number of hops under the delay constraint using a sphere-packing bound.

The research done in this chapter is comparable to previous attempts in which the crucial task of network traffic modeling was looked into. However, it differs in a number of ways. First, our forwarding strategy differs in that sensor nodes can be placed anywhere between the source and the sink along a selected pathway. Second, in terms of energy consumption, our model combines all transmission operations with all of the circuit processing energy consumption. Third, unlike other works, our work in this chapter is based on real-time queuing theory. This theory uses a simple stochastic M/M/1 queuing model to inspect how well a duty-cycle MAC layer performs. Additionally, it simultaneously analyzes its QoS factors in terms of average delay, energy consumption and throughput.

8.3 QoS-Based Multimedia Traffic Modeling Framework

Our proposed data delivery framework is composed of two main components: a retransmission channel access mechanism and duty-cycle modeling.

8.3.1 Analysis of Retransmission Channel Access Schemes

The proposed framework enables a comprehensive comparison of two routing schemes in which medium access to the channel is achieved through a request to send/clear to send/data/acknowledgment (RTS/CTS/DATA/ACK) handshaking to guarantee successful end-to-end retransmission or multi-hop transmission. The transmission schemes in distributed systems can be categorized as

1. Hop-by-hop retransmission routing scheme: in this scheme, at every next hop, the intermediate node checks the correctness of the data packet and requests retransmission with a negative acknowledgment (NACK) packet until a correct data packet arrives. After that, an ACK packet is transmitted to the sender node, indicating a successful transmission. Figure 8.1 depicts the mechanism of retransmission, whereas the first data packet, for example, fails between the nodes 2 and 3. Then node 3 sends a NACK data packet asking node 2 for retransmission. After that, node 2 retransmits the data packet, and node 3 transmits an ACK data packet after successfully receiving the data packet.
2. End-to-end retransmission routing scheme: the intermediate nodes simply forward received data packets to the next hop and do not check the correctness of the data packets until they arrive at the sink. Figure 8.2 depicts the mechanism of a retransmission routing scheme in which data packets forwarded to the next hop are not checked until they arrive at the sink. Moreover, the sink checks the correctness of the data packets and retransmits with a NACK packet to the source if the data packets are incorrect.

8.3.2 Modeling Duty-Cycle Node Operations

We propose a Markov discrete-time stochastic process M/M/1 queuing model so as to meet the QoS requirement of the targeted multimedia sensor application. This proposal is made with a realistic conjecture of a finite queue with the ability to hold m packets for duty cycling, which is controlled by the scheduled operation of the sensor nodes. In the proposed model, the arrival and departure of data packets is

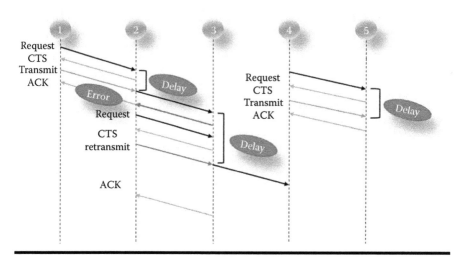

Figure 8.1   Hop-by-hop retransmission scheme.

operated under the assumption that there exists a finite queue size. In addition, the proposed model makes its assumptions based on Reference 16, which is similar to Reference 11, and Reference 11 has been proven to be a perfect example of a realistic scenario. It is evident from the proposed Markov model that we can model the power transition of each sensor node in the network using a discrete-time M/M/1 Markov chain. This characterizes a different predefined status for a node of an event at the wake-up/sleep mode of the duty cycle. Table 8.1 lists all the notations used in the model.

A node may exchange its status, slot by slot, which corresponds to the transition from one state to another in the Markov chain. The proposed Markov model has

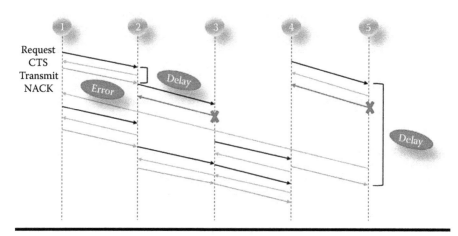

Figure 8.2   End-to-end retransmission scheme.

Table 8.1 Notations of the Proposed Model

Symbol	Definition
$A(t)$	The total number of packets arrival
i,j	The state of none
π_i	The steady state of power for the node of state i
$P_{i,j}$	The probability of transition state I to state j for the node
λ	Expected DATA packet arrival rate at the MAC layer
β	The independent probability of transmitting DATA packet
δ	The time of inter-arrival of packet arrival
m	The number of packets in the queue
M	The capacity of queue for DATA packets in units

limited queuing capacity with finite state slots from left to right, which correspond to a 0 state for processing packets in the queue and so on to m packets in the queue (full queue). Specifically, if a packet arrives and the queue is full, then the packet is simply dropped; nevertheless, the packets are removed from the queue when they are successfully transmitted. By contrast, when the queue is neither full nor empty, then a node may obtain access to the media to transmit packets with an independent probability. The analysis of the Markov discrete-time M/M/1 queuing model offers insights into the traffic behavior of sensor networks in general and points to an idea for a control algorithm. The steady-state probability and the transition probabilities of moving from one state to another can be described as $P_{0,i} = \lambda_i * \delta$, where $i = 0, ..., M$. And since

$$P_{0,M} = \lambda \geq M, \tag{8.1}$$

then

$$P = \beta * \lambda_0 * \delta, \quad \text{for } i = 0, ..., M, \tag{8.2}$$

and

$$P_{i,i-1} = \beta * \delta * \lambda_{j-i+1} * \delta + (1 - \beta * \delta) * \lambda_{j-i} * \delta, \quad \text{for } i = 1, ..., M-1 \tag{8.3}$$

and

$$P_{i,M} = \beta * \delta * \lambda_{\geq M-i+1} * \delta + (1 - \beta * \delta) * \lambda_{\geq M-i} * \delta, \quad \text{for } i = 1, ..., M. \tag{8.4}$$

Therefore,

$$P_{i,j} = 0, \quad \text{for } i = 2, \ldots, M, \quad \text{and} \quad j = 0, \ldots, i - 2. \tag{8.5}$$

Let's assume that the proposed Markov model with a finite set of power mode transition states for each node is defined in space as $S = 0, 1, \ldots, M$ and that the transitional probabilities $\pi_m(t)$ in matrix P are defined as being in state m (the system has m packets). Then, the steady-state equations for each schedule-driven duty-cycle node operation can be described as follows:

$$\pi_m(t) = \pi_{m-1} * (t) * \beta * \delta + \pi_m(t) * (1 - \lambda * \delta - \beta * \delta)$$
$$+ \pi_{m+1}(t) * \beta * \delta + o(\delta), \forall m! = 0. \tag{8.6}$$

The proposed model is considered to be an irreducible, periodic, and recurrent non-null Markov chain; therefore, the model possesses the unique stationary probability $\Pi_m(t) = (\Pi_0(t), \ldots, \Pi_M(t))$, where $\Sigma^M \Pi_m(t) = 1$, $\Pi_m(t) * P = \Pi_m(t)$, which strictly provides that the mean rate of arrivals per state λ is less than the mean rate at which packets are obtained by the server per state β. Moreover, the queue length will become stable, and the number of packets in the queue will be finite under this balanced assumption. Thus, both the packet arrival information λ and the probability of successful transmission β for a specified schedule-driven duty-cycle node operation in MHC become variables in the transition matrix P. Previously, $\Pi_m(t)$ was considered to be a unique stationary probability; hence, $\Pi_m(t)$ can represent a function of both the packet arrival information λ and the probability of successful transmission β.

8.3.3 Energy and Delay Modeling

As stated in Reference 7, energy efficiency still represents the foremost challenge within the design of the future generation of wireless networking (e.g., 5G and IoT) due to the massive number of connected devices. The proposed Markov chain evaluates the energy consumption for multi-hop network communication by defining a critical path loss for considering the randomness of the hop-distance between connected nodes. The main concepts in terms of delay and throughput will be described below. The proposed chain model assumes that every sensor node in the network has the following four main power transition states: transmit, receive, idle, and sleep, which are denoted as P_{TX}, P_{RX}, P_{idle}, and P_{sleep}, respectively. Furthermore, $P_{total}(n)$ is defined as the total power consumption for these four power states during the transmission period. The cost of the energy transition between these four states may be obtained by multiplying the total cost of a single transition by the average number of transitions. However, the total energy consumption at each power state and the energy cost of transitioning from one state to another should not exceed the total energy resource. The energy consumption of a successful hop-by-hop transmission from the source to the sink is given as

$$Energy_{hbh} = \sum_{(i=1)}^{n} \frac{1}{(1 - PER_i)} * E(\Pi_m(t)),$$

(8.7)

where *PER* defines the packet error rate at hop *i* of the selected path in the specified direction.

8.3.4 Throughput Modeling

Throughput is defined as the amount of information successfully delivered within a specified unit of time. The throughput is calculated during the duty-cycle of the node operation on information packets within a given cycle time. The successful transmission of information is described as the probability of sending out information between two or more connected nodes (i.e., *k*) that are competing for media access in the network, which is a function of the current steady state of the power transition mode of the nodes. For MHC, the number of nodes in the network is *N*, the MAC layer DATA packets size is *S*, the length of the cycle is *T*, and $\Pi_m(t)$ is known. Furthermore, the only variable is the probability of successful DATA packet transmission, which can be obtained according to the media access protocol. Thus, the throughput of the network can be determined as

$$Th_{MHC} = N * 1 - \Pi_m(t) * p_s * S/T.$$

(8.8)

8.3.5 Path Loss Model

Propagation phenomena are diverse and complex because of the separation of the receiver and the transmitter, in addition to having a random variation in power due to path shadowing, which causes the signal strength to decay exponentially with respect to the separation. There are various propagation models for path loss, which can be categorized into the following three types: (1) empirical models, (2) deterministic models, and (3) stochastic models. For the free space model, a path loss is proportional to the square of the distance in the IEEE 802.15.4 standard, where $PL(d_0)$ is the path loss at a reference distance d_0 in meters. Existing techniques either consider that the path loss is known before by assuming that the WSN environment is free space or obtain the path loss through extensive channel measurement and modeling by measuring both SNR and distances in the same WSN environment prior to system deployment. However, an accurate knowledge of the path loss is required in order to obtain an accurate estimate of the inter-sensor distance from the corresponding SNR measurement. We defined the mean path loss as a function of the hop distance in relation to the power of the path loss exponent

ξ as $PL(d_i) \propto (d_i/d_0)^{\xi}$. Therefore, our model assumes that there are multi-hop nodes between the source and the sink; thus, the relationship between the path loss and the multi-hop distance may be obtained by

$$PL(d_i) = PL(d_0) + 20\xi * \log_{10}\left(\frac{d_i}{d_0}\right) + \varepsilon(0,\sigma), \quad i = 1, 2, \ldots, n. \tag{8.9}$$

8.3.6 Failure Modeling

We characterize the failure parameters in this domain by exploiting the synergy between fault detection and fault tolerance (FT) in WMSNs. The FT parameters are leveraged based on Markovian models via the coverage factor and the sensor node failure rate.

8.3.6.1 Coverage Factor

The coverage factor c is defined as the probability that the faulty active node in a WMSN paradigm is correctly diagnosed, disconnected, and replaced by a good inactive spare node. The c estimation can be determined by $c = c_k - c_c$, where c_k denotes the accuracy of the fault detection in diagnosing faulty sensor nodes and c_c denotes the probability of an unsuccessful replacement of the identified faulty node with the good spare one. Whereas c_c depends on the sensor node switching circuitry and is usually a constant, c_k's estimation is challenging because different fault detection approaches have different accuracies.

8.3.6.2 Sensor Failure Rate

The failure rate can be represented by exponential distribution with a failure rate of λ_s over the period t_s. The failure rate curve approximation by piecewise exponential distributions is analogous to a curve approximation by piecewise straight line segments. Consequently, the cumulative distribution function (CDF) for the WMSN with an exponentially distributed failure rate can be represented by $F_s = p = 1 - e^{-\lambda t_s}$, where p denotes the cumulative probability of a sensor node failure and t_s signifies the time over which p is specified.

8.4 Use-Case Transmission Modeling

Unlike existing models, our model proposes that the radio propagation model approximates an isotropic property to present the effects of irregularity on the localization technique for correct MAC in cases of asymmetric radio collision between two nodes (which occurs when a node is unable to reserve the wireless channel) and on the routing performance through paths asymmetrical to the traffic behavior in the two transmission schemes. Our model is defined as geodirectional-cast forwarding that consists of a set of mathematical expressions that represent the non-isotropic lognormal path loss with differences in transmit power levels of 0, −3, −5, −7, −10, and −25 dB, as well as IEEE 802.15.4, respectively, according to Equation 8.9, as shown in Figure 8.3. The SNR, in the function of distance on a semi-logarithmic scale on multi-hop sensor nodes, varies according to the propagation direction from the node to its neighbor.

Figure 8.3 was constructed with the average of decreasing SNR values of packets sent at different power levels from sensor nodes to the sinks. The DOI is introduced to denote the irregularity of the radio pattern. The DOI is based on the distance over which a node can hear its neighbor. It is defined as the maximum path loss percentage variation per unit degree change in the direction of radio propagation [17]. To establish a radio irregularity model, our propagation model relies on real data values (which have been repeatedly used in many experiments on MICAZ sensor nodes in vehicle tracking systems) to approximate the radio irregularity by calculating the corresponding DOI value [17]. To reflect the path loss for a specified angle toward an optimal forward direction from the next hop to the nearest neighbor, a new coefficient K is defined; directional forwarding is used to adapt the bounds of the DOI model value between an upper and lower signal propagation for the two categories of retransmission schemes by adjusting the path loss for forwarding to the next nodes that have the best progress toward the sink in the specified direction. Beyond the upper bound, all neighbors are outside transmission range (TR); within the lower bound, all neighbors are guaranteed to be within the inner TR, and the signal is strong enough to be received correctly. Therefore, it becomes more critical to carefully select the sensor nodes that participate in the sensor transmission range,

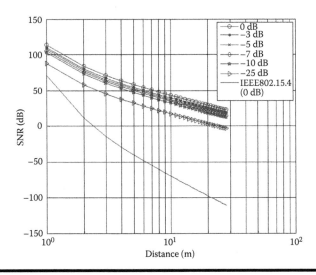

Figure 8.3   SNR multipath fading for MHC at different power levels.

forwarding the information against its resource consumption. Thus, the DOI modeling results are given as

$$SNR = P_{TX} - DOIAdjustedPL(d_i) + fading, \tag{8.10}$$

where DOI-adjusted $PL(d_i) = PL(d_i) * K_j$, and, K_j is defined as the adjustable jth coefficient used to adjust the path loss value according to a specified direction. Thus, minimizing the amount of energy consumption and adjusting the transmission range as much as possible through observed DOI values can significantly prolong the lifetime of the sensor network.

8.5 Performance Evaluation

In this section, realistic sensor nodes and network specifications are adopted. The MICAZ sensor characteristics and network topology scenario were defined with the following assumptions under the IoT application behavior:

1. Each sensor node in the network is represented by a tuple $Sn = <V, E, P_v, PE>$ where V and E specify the network topology with the sensor nodes V and link connectivity E. P_v is a set of functions that characterizes the properties of each link (e.g., the link quality).

2. The construction of the network topology depends on the location management, which determines the localized information of uniformly deployed sensor nodes. It assumes that all MICAZ sensor nodes are in a fixed position and that the sink is at the origin (0, 0). For MHC, there is at least a single path connecting the source to the sink through intermediate sensor nodes, issuing network commands such as "sleep", "idle", and "wake up"; changing the level of transmit power; and synchronizing the transmission time.

3. Vehicles are simulated as sources of abstract sensor readings that can be detected in sensing range of the sensor node. Each reading is dispersed in the simulation over a certain point and at a certain time as

$$Y(p,t) = \sum_{allvehicles_i} \frac{(Y_i(t))}{(K * d_i(t) + 1)^a}, \tag{8.11}$$

where $Y(p, t)$ denotes the position of the vehicle physical process at a certain point p and at a certain time t; $Y_i(t)$ denotes the position of the ith vehicle at time t and is called a snapshot of vehicle, which helps to define the maximum possible number of pickups of snapshots for all vehicle s; and $d_i(t)$ denotes the distance of the ith vehicle from the sensor node at time t in the sensing range.

4. Each node tries to access the channel through a handshaking technique, which is designed to resolve hidden and exposed terminal problems. However, when the source node generates DATA packets for transmission, it sends an RTS message to its neighbor, which responds with a CTS message.

5. The proposed model depends on Lee's model calculation [12] to estimate the path loss because it is the most commonly used in urban areas. The expected hop distance can be attributed to the influence of the path loss due to the various propagation directions, which is a function of SNR for different power values P_{TX}, as shown in Equation 8.10.

8.5.1 Simulation Results

This section analyzes the QoS parameters obtained from the analysis of the effects of radio irregularity on a multi-hop network routing protocol with a focus on realistic scenarios at low densities considering effects that might have a significant influence on the performance of geographic routing. In the following subsection, the average packet delay, the energy consumption for transmission, and throughput are analyzed in view of the impact of radio irregularity on the neighbor-discovery routing technique of both the hop-by-hop and end-to-end retransmission schemes.

8.5.1.1 The Impact of Radio Irregularity on Energy Consumption

We analyze the assumed reduction of energy consumption, which is taken as a key issue, in terms of reducing the power overhead through a new approach for MHC when $TR > D_{max}$, and hence, when the number of hops is signified as the upper limit, MHC will become more favorable. Subsequently, we must analyze the power overhead in retransmission schemes. Figure 8.4 depicts the amount of energy consumed when the overhead power is reduced on each node. We use the discrete

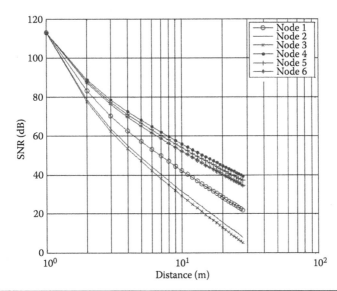

Figure 8.4   SNR vs. distance for MHC at different power levels.

Table 8.2 Simulation Parameters

Parameter	Value
P_{TX}	59.1 mW
P_{RX}	52.2 mW
P_{idle}	55.0 mW
P_{cir}	12.0 mW
$P_{startup}$	1.0 mW
Preamble length	4 bytes
Frame length	8 bytes
Wave length	0.1224489796 bytes
Path loss exponent	1–6
Noise floor	−115.0 dBm
Modulation	Non
Coding	NRZ & Manchester
Link capacity	19.2 kbps

M/M/1 Markov queuing model, which matches the power overhead in sending and receiving control messages within a sensor network. The results shown are for the hop distances between nodes for the two retransmission schemes with different packet error rates (PERs) (Table 8.2).

The model is useful for most real-time WSN applications, particularly when the traffic load is heavy and changes over time. When the PER is very low, the energy consumption increases because the power amplifier must consume more energy to guarantee a smaller PER based on Equation 8.7.

8.5.1.2 The Impact of Radio Irregularity on Average Delay

Figure 8.5 depicts the average delay versus the hop distance from the source to the sink with different PERs for the two retransmission systems. From the figure, we can observe that the average delay starts by hovering over zero before increasing as the hop distance increases. Moreover, the maximum delay in hop-to-hop retransmission with non–return-to-zero (NRZ) coding is greater than the maximum delay in end-to-end retransmission using the same coding. When the number of hops increases to above two, the average maximum delay in the end-to-end retransmission system becomes 31.9%, which is still less than that of the hop-by-hop retransmission scheme. When the number of hops increases, the average delay time for the

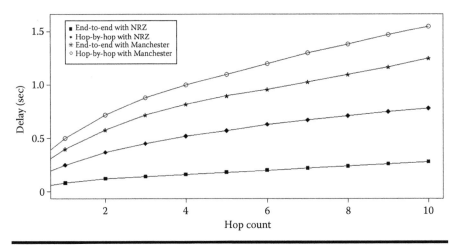

Figure 8.5   The average delay in retransmission schemes.

end-to-end system improves to 41.8%, compared to the improvement gained in the hop-by-hop system of 47.9%. This is because in hop-by-hop systems, every interme-diate node must transmit an ACK/NACK packet, which generates more traffic and increased delay. Manchester encoding caused longer delays compared to the NRZ encoding because the packet length is double-decoded. Hence, the average delay increases with the number of hops, because β and Π_m are subtler to an increase in the number of hops. This can be due to path loss caused by different propagation directions.

8.6 Concluding Remarks

The main question for a multi-hop IoT system such as an ad hoc intelligence trans-portation system (ITS) network is whether it is more beneficial to route over many small hops or fewer longer hops. One of the advantages of short hops is the SNR gains for getting the furthest neighbor that can be used for adequate reliability. Consequently, this chapter investigates the claim that effective anisotropic radio properties inside the deployment area has a role to play in the energy consumption and traffic system model. We developed a traffic model system by using the Markov discrete-time M/M/1 queuing model for duty-cycle nodes so as to show the perfor-mance of two different retransmission systems and to stem the fixed distribution of the probability of packet transmission for the two systems so as to look into the performance of the duty cycle of the MAC layer in concurrently examining the delay, throughput, and energy consumption.

QoS specification can be used to provide various priorities to guarantee a certain level of performance to a data flow in accordance with requests from the application. However, QoS is considered a must in WMSNs, especially when the network bandwidth capacity is limited. Therefore, WMSNs offering guaranteed QoS for multimedia traffic should be flexible enough to support different application-specific QoS requirements (such as energy efficiency, end-to-end delay, and throughput levels) in a heterogeneous setup. Nevertheless, available WMSNs are not adequate yet to achieve QoS requirements in the IoT era. In addition, QoS modeling becomes more significant in dense deployment of WMSNs, but little attention was paid to this in the literature, and more traffic modeling attempts are necessary in this direction. In order to fulfill the QoS requirements for each user, an optimized and efficient radio resource management solution is required as well.

References

1. E. David, N. Ndih, and S. Cherkaoui, On enhancing technology coexistence in the IoT Era: ZigBee and 802.11 case, *IEEE Access*, vol. 4, pp. 1835–1844, 2016.
2. G. Solmaz, M.I. Akbas, and D. Turgut, A mobility model of theme park visitors, *IEEE Transactions on Mobile Computing*, vol. 14, no. 12, pp. 2406–2418, 2015.
3. G. Solmaz and D. Turgut, Tracking pedestrians and emergent events in disaster areas, *Journal of Network and Computer Applications*, vol. 84, pp. 55–67, 2017.
4. W. Ibrahim, A. Taha, and H. Hassanein, Using smart vehicles for localizing isolated things, *Computer Communications*, vol. 74, pp. 16–25, 2016.
5. F. Al-Turjman, H. Hassanein, and M. Ibnkahla, Towards prolonged lifetime for deployed WSNs in outdoor environment monitoring, *Ad Hoc Networks*, vol. 24, pp. 172–185, 2015.
6. T. Soyata, L. Copeland, and W. Heinzelman, RF energy harvesting for embedded systems: A survey of tradeoffs and methodology, *IEEE Circuits and Systems Magazine*, vol. 16, no. 1, pp. 22–57, 2016.
7. S. Oteafy and H. Hassanein, Resilient IoT architectures over dynamic sensor networks with adaptive components, *IEEE Internet of Things Journal*, vol. 4, no. 2, pp. 474–483, 2017.
8. Libelium - Connecting Sensors to the Cloud, *Libelium.com*, 2017. [Online]. Available at: http://www.libelium.com (Accessed July 27, 2017).
9. M. Hasan and F. Al-Turjman, Evaluation of a duty-cycled asynchronous X-MAC protocol for vehicular sensor networks, *EURASIP Journal on Wireless Communications and Networking*, vol. 1, no. 1, 2017.
10. M. Sikora, J.N. Laneman, M. Haenggi, D.J. Costello Jr, and T. Fuja, On the optimum number of hops in linear wireless networks, in *Proceedings of the IEEE Information Theory Workshop*, pp. 165–169, 2004.
11. D. Jung, T. Teixeira, and A. Savvides, Sensor node lifetime analysis: Models and tools, *ACM Transactions on Sensor Networks (TOSN)*, vol. 5, no. 3, pp. 1–29, 2009.
12. W.C.Y. Lee, Lee's model cellular radio path loss prediction, in *IEEE 42nd Vehicular Technology Conference*, pp. 343–348, 1992.
13. M. Watfa, S. Selman, and H. Denkilkian, A battery-aware high-throughput MAC layer protocol in sensor networks, *International Journal of Distributed Sensor Networks*, vol. 6, no. 1, pp. 259–269, 2010.

14. A. Antonopoulos and C. Verikoukis, Network-coding-based cooperative ARQ medium access control protocol for wireless sensor networks, *International Journal of Distributed Sensor Networks*, vol. 8, no. 1, pp. 601–621, 2012.

15. I. Stanojev, O. Simeone, Y. Bar-Ness, and C. Myeon-gyun, On the optimal number of hops in linear wireless ad hoc networks with hybrid ARQ, *6th International Symposium on Modeling and Optimization in Mobile, Ad Hoc, and Wireless Networks and Workshops*, pp. 369–374, 2008.

16. D.P. Bertsekas, R.G. Gallager, and P. Humblet, *Data Networks*, vol. 2. Prentice-Hall: Englewood Cliffs, NJ, 1987, pp. 363–371.

17. G. Zhou, T. He, S. Krishnamurthy, and J.A. Stankovic, Impact of radio irregularity on wireless sensor networks, in *Proceedings of the 2nd International Conference on Mobile Systems, Applications, and Services*, pp. 125–138, 2004.

Chapter 9

Information-Centric Framework for the IoT: Traffic Modeling and Optimization

9.1 Introduction

The information-centric Internet of things (IoT) is an approach that propels the Internet away from the host-centric paradigm, which is based on perpetual connectivity and end-to-end principles, to an ad hoc network architecture in which the focal point is the provided information and data content is treated as a first-class entity in network architecture [1]. In this ad hoc network, wireless devices communicate without a fixed infrastructure. It usually consists of equal-capability nodes that communicate wirelessly, evading a central command/data request [2,3]. Efficient content distribution is one of the main motivations behind this IoT paradigm. However, the authors in Reference 4 argue that it is not enough to motivate a shift to a new network infrastructure. Hence, in Reference 5, the authors expound on some other additional advantages (such as persistence, unique naming, security, and disruption tolerance, among others) that makes IoT stand out. Moreover, the security in the current host-centric model protects the channel between the server and client through a transport security layer (TSL). This requires the client to trust the server to deliver the correct information over the channel. In contrast, the IoT provides name-data integrity and origin verification of the requested content/information, independent of immediate sources [5]. Furthermore, the model enables ubiquitous caching with retained name-data integrity and authenticity, something the current Internet communication model does not provide [5]. By using persistence and unique naming of contents, the IoT approach overcomes the problems of name-object binding breakage, which is experienced by the current host-centric networks due to its object locator nature [5]. Because of the host-based nature of the current system, mobility and multihoming become a problem for managing end-to-end connections and determining which path to choose for these connections. The IoT system does not have end-to-end connections; hence, it eliminates this limitation altogether. In addition, switching to an information-centric paradigm in connecting objects/things can potentially save energy because data is transported, on average, over shorter distances while distributing the content copies everywhere. However, it raises further overhead issues in terms of cache coordination and global management of the in-network storage resources [6].

Looking at the above-mentioned motivations, we can conclude that the IoT can be considered as the future ad hoc network of the current Internet. Some other motivations include scalability and cost-efficient content distribution and disruption tolerance [7]. Nonetheless, to make the IoT a more desirable architecture, there are still some challenges that need to be overcome. For instance, the authors in Reference 8 point out that when caching takes place in the network, several traffics compete for the same cache. Hence, caching is one of the challenges that must be looked at. Quality of service (QoS) is another important part of any ad hoc IoT architecture [8]. However, only a few of them provide details about useful QoS mechanisms, whereas the rest treat this issue casually. One example of this is knowing that the experienced delay in IoT networks can be one of the most critical/challenging QoS factors to be considered and addressed. Moreover, mobility and data relaying can be a key feature in any ad hoc IoT paradigm, which can theatrically affect the estimated system performance in practice [9]. It is worth pointing out also that when the original data/information resources, such as regular servers and network storage, are considered as data sources in the IoT, their access and reachable data paths become profoundly overwhelmed to a level that can intensely degrade their performance. Additionally, there will be a noteworthy exhaustion of their energy resources caused by the mobility of wireless devices and connectivity disruption in the IoT network due to continuously changing locations.

Hence, it is essential to dismiss the publishers from their heavy responsibility, which is expected to be augmented by the consumers in a typical IoT network, by stressing the notion of data repeaters. In References 10 and 11, the authors point out that to decide on an effective resource allocation in accessing and distributing the IoT contents, replication points need better fitting and more accurate traffic characterization models to be used. Although there are several models for network characterization (such as Poisson distribution models, Wavelet-based models, Markov models, and others [12–14]), these models only describe traffic characteristics using randomly assumed parameters, such as the Poisson arrival, rather than depending on a realistic feature in the network. Besides, caching alone is not enough for the demands of Internet applications at peak times, during which data demands come heavily from specific hot spots/regions of the network. Moving from today's host-centric trusted model to the information-centric one represents a key challenge in terms of security and has to be adequately addressed [6,15,16].

In this chapter, we propose a framework that can approximate the IoT data requests by using ellipses because of the observed cyclic and nonsymmetrical properties in the Internet traffic. The content demands are discerned in statistics and in a traffic mapped–geometric model called content demand ellipse (CDE). The CDE-based framework incorporates new secure/safe entities between the data publishers and consumers called information repeaters (IRs). These IRs provide much faster data access, redistributing data flow among the different demanding regions of the network and relieving the main data publishers from expensive overloads. Furthermore, they provide better opportunities for cache coordination and global management of in-network storage resources usage. For practical conclusions, we used real network traffic to get the CDEs. And for the imbalanced demands, some resources have been reallocated by using an integer linear program (ILP) optimization approach. Accordingly, the main contributions of this article can be summarized as follows:

- The proposed CDE-based framework incorporates a multi-tier architecture that introduces new IoT components between publishers and data users, namely the IRs. These IRs are not simply caches/replication points. They are other equivalent resources to the original data publisher/source, such as redundant cameras, sensors, servers, and/or network-attached storages. They are introduced to relieve the original data publisher from any undesired overhead.
- A novel IoT-specific ILP is proposed to reallocate static/mobile IRs for better overhead release from the original publisher and better content accessibility. This ILP is significant in providing the most appropriate resource in case of irregular data request patterns while maintaining a limited delay.
- A novel traffic analyzer in IoT, obtained based on the CDE, is proposed and employed by the ILP in order to consider the most accurate network status description while suggesting the resource reallocation.
- Moreover, this work verifies that in an IoT network with randomly distributed redundant/storage resources, namely the IRs, a subset of these IRs can be more appropriately reallocated to meet key end user and publisher requirements, even under the heavy traffic conditions in IoT applications.

The remainder of this chapter is organized as follows. Section 9.2 surveys the related work. Section 9.3 highlights the main assumptions and system models. Section 9.4 describes the proposed CDE-based framework. Section 9.5 explains the targeted case studies in IoT. In Section 9.6, the effects of the varying network traffic on the proposed CDE framework have been investigated. Finally, Section 9.7 concludes this research.

9.2 Related Work

The current Internet paradigm is in danger of being overwhelmed by the rapidly growing number of users. Unfortunately, the Internet in use today does not have the capabilities of handling a high number of users' requests. Consequently, the described IoT paradigm in this article aims at handling a colossal amount of user requests, revolutionizing the future of Internet surfing [3]. One major advantage of the IoT is the ability to eliminate common overhead communications unrelated to their physical location or their hosting machines by assuming a different network architecture that has the ability to respond differently to changing network conditions and available resources. In the targeted IoT paradigm, entities consist of communicating nodes and caches connected to nodes, and data items are assumed [1,17]. A node can either be a publisher, a user, or an intermediary node. A publisher can be defined as an entity with the ability of fulfilling user requests and providing certain information. One other important feature of the information-centric Internet is the ability to publish only once and then have copies/replicas placed/republished at different locations of the network [5]. The authors in References 18 and 19 conclude that this feature reduces the response time for numerous data requests and also reduces the distance traveled by a request between the producer and consumer in case the data has been corrupted on its way to the user. This feature provides the ability for the proposed hierarchical IoT paradigm to reproduce the exchanged data that is only produced once by the original publisher at different/multiple intermediate locations/IRs between the user and the original data publisher. Because clean and correct data has to be delivered to the user, the IRs in the IoT provide/recover the corrupted data without the need to go back all the way to the original publisher, and hence, the distance is significantly reduced. In order to apply this feature, caching approaches are usually performed over the intermediate nodes of the network. This can be implemented by either opportunistically exploiting data repeaters over the targeted network or by caching at intermediate nodes in order to provide in-network replicas.

For example, the authors in Reference 20 aim at caching the most popular data at intermediate nodes in order to increase the ratio. However, this approach is not always effective due to varying popularity metrics from one user to another. In References 21 and 22, the authors argue that caching solutions at the edge of the network can be more effective in placing the requested data near the end user. This can partially solve our targeted problem, as there will be no replicas available for in-network processing, which is a typical scenario in the IoT. In Reference 23, the authors propose the distribution of caching helpers all over the network. Nevertheless, it is caching only video streams over powerful cellular networks, which is not always the case in energy-limited IoT networks.

On the other hand, the authors in Reference 24 designed and implemented a SoftRepeater, which is a Wi-Fi-compatible software solution, in which high-rate stations behave as repeaters for low-rate ones when it is beneficial to do so. Similar proposals are also provided in References 25 and 26. In addition, repeaters can be implemented by populating extra nodes as discussed in References 27 and 28. For example, an information-centric architecture is proposed in Reference 26 as an overlay network based on populating additional infrastructure inside the access networks, called overlay access routers (OARs). These OARs establish an overlay routing substrate for the forwarding of data based on information identifiers. This scheme does not address core requesting issues such as the temporarily and highly demanding regions. Their repeaters' allocation and other network design aspects such as repeaters count and distribution are not considered either.

Our approach divides the load among all the main components of the proposed IoT architecture components, including the users, in order to provide better data cost from the user's perspective according to current network status, such as the available publishers' count, their traffic load, and their distribution. However, serving a multitude of applications/users with varying data request types and locations can significantly degrade the effectiveness of caching-based solutions. Moreover, the additional message exchange overhead cannot scale easily to a growing, dynamic IoT and yet claim inefficiency [29].

Therefore, in this work, a dynamic IR reallocation strategy is proposed to manage the exchanged IoT requests [3]. A CDE-based framework is introduced here based not only on providing data accessibility but also on reallocating available resources based on the requested content popularity and peak demand, which can be determined via an elliptical network traffic model. In general, the proposed approach aims at predicting candidate IRs to meet the requested content while considering irregular network loads. It is expected that this approach could result in more disruption tolerance and on-time data delivery in next-generation networks such as the IoT.

9.3 System Models

In this section, we describe the considered IoT model, network disruption model, and data request/content representation.

9.3.1 Network Model

Users asking for the data content and publishers generating and providing this content to users over the Internet form the backbone of the considered IoT network. Examples for user nodes in this study are things such as a hand-held smartphone, laptop, smart car, etc. Meanwhile, publishers can be represented by nodes with the capability of publishing data contents. Publishers can either be static or mobile. The role of users is exclusively a consumer role, whereas publishers exclusively play a producer role [16,30].

In Reference 27, the authors further argue that intermediary nodes in the network simplify the communication between producers (publishers) and consumers (users) by providing routing and caching services; thus, they are considered in our proposed IoT model. In addition, a new component (the IRs) that represents the available redundant network resources for any specific content to be republished is proposed. These IRs can be any redundant data source (e.g., in-network cache or smart device) or an extra dedicated device that is added by the network operator for repeating the originally generated data at the edge of the network in order to make it closer to the end user. And in this case, it acts like a signal repeater in wireless communication systems. Finally, the gateways (GWs) are in-network devices that connect the sub-network to the Internet. If necessary, they can act as floating IRs after concluding the current network status. These components form the proposed architecture for the targeted IoT paradigm, as depicted in Figure 9.1. According to the description above, IRs are introduced to relieve publishers from any unnecessary transmission/republishing load. The above-mentioned components are constructing a graph that represents the overall IoT. Mainly, this representation can be used to predict future resource allocations.

9.3.2 Delay and Disruption Model

When a content publisher is disconnected from the network, it causes a service disruption and undesired delays in the IoT. It can cause a substantial increase in the exchanged data traffic. As a result of the compact nature of most IoT network components, there might exist a long multi-hop path comprising diverse devices between the data publisher and the end user [12]. Because these devices might not have direct access to the next hop due to mobility and/or energy issues, further delay can be experienced while holding the data until next hop is available [31].

Hence, our objective is to maintain a limit on the maximum (worst) delay that can be experienced between the proposed IoT components by reallocating a subset of the available resources (or IRs). Therefore, the delay model is introduced in Reference 32. Since we are relying on an ILP-based approach, we consider its discretized delay metric that can be tuned to achieve any desired accuracy. Due to dense network topologies formulated in IoT, a relatively long multi-hop path can easily exist between the publisher and the corresponding IR. The delay components we consider in this chapter are the transmission/processing delay ψ (modeled by the number of hops multiplied by ψ) and the propagation delay (modeled based on the speed of signal and the Euclidian distance between the two ends). The latter delay varies based on the utilized technology and its corresponding standards and the transmission medium in the utilized IoT subnetwork.

We define a delay step ω, which is the distance a data request would travel in one time unit. Let E_{ij} be the Euclidian distance between a source node i and a destination node j; then, the discrete propagation delay over a single-hop link (i, j) would be E_{ij}/ω. Hence, the discrete delay over a multi-hop path is the total of the discrete delays of single-hop links that constitute that path. For the sake of our approach generality, a single-hop delay ($D_{single-hop}$) can be defined as $D_{single-hop} = (E_{ij}/\omega) + \psi$, and the total multi-hop delay is simply the sum of all experienced individual hops' delays.

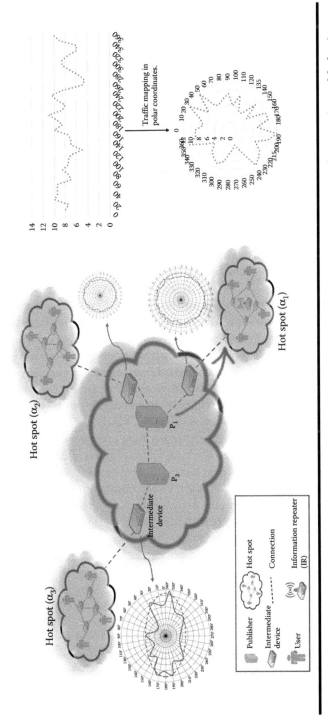

Figure 9.1 Proposed IoT architecture that maps cyclic network traffic in polar system based on the experienced behavior per hot spot/region.

9.3.3 Traffic Representation Model

In order to precisely estimate the required resources/IRs in this study, a practical content representation model that relies on practical assumptions about the exchanged data traffic nature between the communicating entities of the assumed IoT paradigm has to be considered. Key properties to be considered in the Internet traffic are the observed nonsymmetrical and cyclic nature of the requested data [33]. In order to elaborate more on these properties, we present a real traffic data set, which has been used in several studies [33,34], as an example. The used data set targets a backbone network located in North America, and the traffic data contain averages over 5-minute intervals from March 1st to September 11th. Table 9.1 shows one example subset matrix of the observed traffic in practice from that period on April 15th between 16:25 and 16:30, along with row and column totals.

Apparently, we can note that the matrix in Table 9.1 is nonsymmetrical and that there are quite a range of values. Also, it is remarkable that the matrix diagonals are not zeros, as there is some measured traffic that arrives and departs at the same point of presence. For a more comprehensive overview, a time series view of the referenced/used data set is shown in Figure 9.2. The top figure (Figure 9.2a) shows the total traffic during each time interval for the whole data set, and the bottom figure (Figure 9.2b) shows the first two weeks of the data, overlaid in order to show the cyclic nature of the data traffic.

The request for information from the Internet has been observed to be cyclic by nature [33,34]. In our proposed method, we first obtain the period of the cycle in which the IoT traffic is supposed to repeat its pattern, and then, from that period, the frequency of the request is mapped to a polar system. A polar system is a two-dimensional coordinate system in which any point is given by the distance from a reference point and the angle from a reference direction. In this system, coordinated data traffic points form an oval shape called an ellipse. Thus, the ellipse in our study represents the data traffic pattern/behavior per cycle. This ellipse has a number of parameters (such as the semimajor and semiminor axis length, the orientation angle, and the eccentricity parameter) that have been used in quantifying/measuring the targeted IoT traffic in this study. In Table 9.2, a brief definition per parameter with an illustrative diagram in Figure 9.3 is presented.

In this study, we use the previously described ellipse properties/parameters in mapping the observed data traffic in the polar system as shown in Figure 9.1. The time unit in this mapping is flexible (i.e., it can be days, hours, or minutes). In such a resource allocation criterion, the start and stop time are not relevant until irregular traffic is observed, which is the start of a new pattern in the observed traffic. If the traffic flow does not perform in the expected boundaries, then the shape of the ellipse changes, which in turn raises an alarm to the network administrator.

Nonetheless, we cannot rely on the mapped polar coordinates in Figure 9.1 unless it has been simplified by using the above-mentioned ellipse parameters.

Table 9.1 A Five-Minute Traffic Matrix on April 15th, 2014, from 16:25–16:30, in MBPS

Destination													
Source	1	2	3	4	5	6	7	8	9	10	11	12	Row Total
1	0.07	0.07	0.43	0.00	0.06	0.12	0.06	0.00	0.05	0.00	0.00	0.25	1.12
2	0.00	4.09	6.42	0.06	7.07	4.42	1.59	0.02	3.24	0.03	0.16	11.09	38.18
3	0.00	4.70	25.48	4.11	13.99	11.53	3.31	87.27	5.22	0.01	0.08	7.70	163.38
4	0.00	1.93	10.25	1.68	5.63	6.11	2.59	0.01	4.11	2.60	0.04	5.92	40.88
5	0	4.76	0.25	0.01	24.06	0.04	0.01	0.02	1.24	0.02	0.03	18.05	48.49
6	0.00	2.87	32.73	1.55	13.53	4.78	2.89	0.01	9.45	0.08	0.50	7.64	67.02
7	0.00	0.67	4.79	1.92	3.50	2.24	1.25	0.00	0.93	0.02	0.03	3.31	18.67
8	0.00	4.18	2.58	5.80	26.35	0.17	0.16	1.41	10.88	2.11	3.64	16.67	73.97
9	0.00	8.61	12.34	5.71	18.21	11.05	3.84	0.41	36.36	0.02	0.52	17.31	114.37
10	0.00	0.18	0.04	1.71	1.69	0.00	0.06	5.61	0.96	1.82	8.44	0.36	20.86
11	0.00	3.47	3.28	0.54	8.60	0.13	0.93	3.92	1.77	0.81	0.61	2.32	26.38
12	0.00	18.20	16.04	0.83	34.03	11.18	5.64	0.09	25.57	0.08	0.80	47.02	159.47
Column Total	0.07	53.74	105.61	23.94	156.73	52.76	22.34	98.77	99.77	7.59	14.84	137.65	772.80

Source: Internet traffic dataset [online]: http://www.maths.adelaide.edu.au/matthew.roughan/project/traffic_matrix/

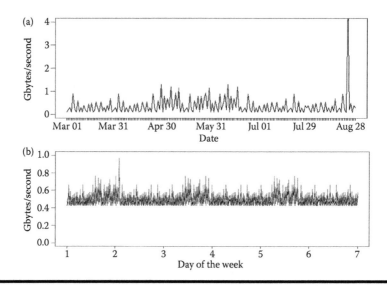

Figure 9.2   The Internet traffic data set in practice [34]; (a) five-minute totals of traffic matrix and (b) Internet traffic for weeks starting March 1st and March 8th.

As shown in Figure 9.4, an ellipse of the best fit is attained. The key parameters that define this best-fit ellipse are the length of the semimajor and semiminor axes and the orientation angle (φ). Additionally, the center of the fitted ellipse (x_0, y_0) and its bias from the origin (0, 0) of the polar coordinates is also critical to the proposed model. The Euclidian distance (d) from the origin given by $d = \sqrt{x_0^2 + y_0^2}$ is significantly affected by the data requests' count. And thus, we use this Euclidian distance (d) to represent the level of traffic demand in the network, in which min(d) is the minimum demand, and max(d) is the maximum demand encountered.

9.3.4 Problem Statement

For the above-described IoT paradigm, including its three main components (the data publishers, IRs, and users' nodes), we aim at *"Prioritizing and determining the minimum number and location of IRs such that; (a) the total cost for the delivered content is minimized, (b) the maximum traffic is offloaded from the original publisher, and (c) all requested demands are satisfied within a limited delay."*

In other words, the main objective is to relieve the original data publishers from extreme traffic demands at peak times while better utilizing the redundant network resources (or IRs) because we want to minimize the number of IRs in order to minimize the total cost of the network. In the following section, we identify a strategy for traffic demands modeling and IRs' reallocation/positioning in the aforementioned IoT paradigm.

Table 9.2 Ellipse Parameters and Description from the Network Traffic Perspective

Parameter	*Definition*
Semimajor axis length	It represents the distance from the ellipse center to the farthest point on the circumference, as depicted in Figure 9.3.
Semiminor axis length	It represents the distance from the ellipse center to the closest point on the circumference, as depicted in Figure 9.3.
Orientation angle (φ)	An angle between the horizontal polar axis and the semimajor axis of the ellipse. It represents the time at which the peak traffic occurs. If the value of the orientation angle is positive, the rotation of the ellipse is counterclockwise. If the value of the orientation angle is negative, the rotation of the ellipse is clockwise.
Eccentricity (e)	A measure of how "*out of round*" an ellipse is. It is given by the formula $e = c/a$, where c is the distance from the center to a focus and a is the distance from that focus to a vertex (see Figure 9.3). It can take a value between 0 and 1. When the value of e is equal to 0, the shape of the ellipse is a circle. When the value of e is 1, the ellipse looks like a line (i.e., it is completely squashed). Note that e is the deviation of the ellipse from a circle-like shape. And hence, when the value of e approaches 1, the traffic is at its steady state; however, when it approaches 0, the traffic becomes unstable and unpredictable. Practically, this parameter represents how much the traffic fluctuates above or below the steady state due to several devices entering/leaving the IoT network at that time period.

9.4 CDE-Based Framework

In this section, we discuss the proposed CDE-based framework, which is recommended to determine the ideal positions of IRs in a given ad hoc IoT network. In this case, limited contending hot spots are prioritized for the IR reallocation process according to the aforementioned ellipse-based traffic model. Through this model, a major parameter that represents the network status is utilized in the CDE framework. In this framework, Algorithm 9.1 is used to show the necessary steps needed to find the content demand ellipse. The algorithm inputs are simply a network graph for the targeted IoT network and a record of previously requested

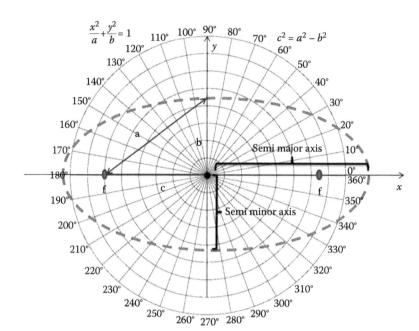

Figure 9.3   Eccentricity of the ellipse is c/a.

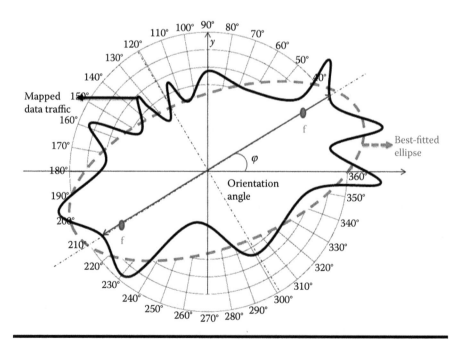

Figure 9.4   Data traffic representation in IoT using the best-fitted ellipse.

demands from the content publisher per time unit. This algorithm finds the content demand ellipse and returns back its major parameters, which are the eccentricity (e), orientation angle (φ), and the Euclidian (d). With these parameters, the system is programmed to respond to any anomalies, choose whether a given IR needs to be reallocated, and forward its data to the next IR in order to avoid any unnecessary traffic in the network.

Algorithm 9.1: Content Demand Ellipse (CDE)

---

1. **Function Traffic_Characterization** (G,S)
2. **Inputs**
3. G: Targeted IoT network graph with n nodes.
4. S: Set of all requests in time period T.
5. **Outputs**
6. $\vec{e}$: Eccentricity vector of the best-fitted ellipses,
7. $\vec{\varphi}$: Orientation angle vector of the best-fitted ellipses,
8. $\vec{d}$: Euclidian distance from the polar origin vector of best-fitted ellipses,
9. **Begin**
10. **For all** $GW_i \in G$
11.   Map network requests (S) onto polar coordinate system to obtain closed curve.
12.   Fit an ellipse to the closed curve
13.   Compute e_i, d_i, and φ_i values the corresponding to ellipse
14. **End**
15. **Return** $\vec{e} = \{e_1, e_2, ..., e_n\}$, $\vec{\varphi} = \{\varphi_1, \varphi_2, ..., \varphi_n\}$
 $$\vec{d} = \{d_1, d_2, ..., d_n\}$$

As we previously mentioned, the d parameter is an indicator for imbalanced network traffic. It represents the distance from the center of the fitted ellipses to the origin of the polar coordinates. The higher the value, the more resources should be reallocated when the content demands an increase. As for the orientation angle, it gives the time when the peak occurs. This is detailed in the following discussion for each case. Accordingly, the total of available IRs is considered, and an ILP is formulated to minimize the cost. The problem is casted as a constrained optimization problem with the objective of minimizing the IoT network cost (C) while maintaining a certain level of the publishers' load, maximum delay, and limited data repeaters. By using the optimization variables defined in Table 9.3, the total data delivered per time unit through the IoT publishers can be expressed as

$$C = \sum_{j=1}^{k}\sum_{i\in\mu_j} l_i P_i + d + \sum_{i=1}^{n} R(i) + \sum_{i=1}^{n} R_{IR}(i) + \sum_{i=1}^{n} \max\{R(i)\} + D_{\max}, \quad (9.1)$$

Table 9.3 Summary of Important Parameters in CDEs

Notation	Description
r	Total count of candidate regions for IRs.
l_i	Percentage (portion) of the publisher load that can be repeated by an IR.
P_i	Publishing load achieved by data publisher i.
α_i	$\begin{cases} 1, & \text{when repeater is placed at } i \\ 0, & \text{otherwise} \end{cases}$
γ_i	Peak demands at region i.
K	Set of all data publishers in an IoT.
k	Total number of data publishers in an IoT.
N	Available IRs.
β_j	Load percentage of a publisher j that is consumed by the noncandidate regions.
μ_j	Set of all candidate regions in the vicinity of a publisher j.
R	Set of all candidate regions for *IRs*
f_{ij}	The data unit (content) flow from ith publisher to jth IR.
D_{max}	Maximum allowed delay.
l_{ij}	The data unit (content) flow from ith IR to jth end-user device.
t_i	The traffic capacity of a publisher i.
T_i	The traffic capacity of an IR i (i.e., maximum data units that can be republished by ith IR).
$M(i)$	A set of indices such that $j \hat{I} M(i)$ if node j is reachable by a publisher i.
$N(i)$	A set of indices such that $j \hat{I} N(i)$ if node j is reachable by an IR i.

where the cost of delivering the data is assumed to be proportional to the amount of data transmitted.

In Equation 9.1, i represents the individual data points in a search space, $R(i)$ is the resource request at the ith time unit, and $R_{IR}(i)$ represents the resource allocation at the ith time period obtained from a rectified curve of the best-fitted ellipse of the IR. Moreover, Equation 9.1 gives the formula to calculate the total resource allocations for the experienced demands, baseline, and dynamically allocated IRs, and thus, the total savings can be computed as follows:

$$Savings\,(\%) = \frac{Baseline\text{-}allocated\ IRs}{Baseline} \times 100. \qquad (9.2)$$

That being said, the targeted optimization problem can be formulated as

$$\min C, \qquad (9.3)$$

which is subject to

$$\sum_{i=1}^{r} \alpha_i \leq N, \qquad (9.4)$$

$$\sum_{i \in \mu_j} l_i \leq 1 - \beta_j, \quad \forall j \in K, \qquad (9.5)$$

$$l_i P_i + \alpha_i \gamma_i \geq \gamma_i, \quad \forall i \in R, \qquad (9.6)$$

$$0 \leq l_i \leq 1, \qquad (9.7)$$

$$\sum_{j \in N(i)} D_{single\text{-}hop} \cdot f_{ij} + \sum_{j \in M(k)} D_{single\text{-}hop} \cdot l_{kj} \leq D_{max}, \quad \forall 1 \leq i \leq n, \ 1 \leq k \leq N, \qquad (9.8)$$

$$\sum_{j \in N(i)} f_{ij} \leq t_i, \quad 1 \leq i \leq N, \qquad (9.9)$$

and

$$\sum_{j \in N(i)} f_{ij} + \sum_{j \in M(i)} l_{ij} \leq T_i, \quad 1 \leq i \leq N. \qquad (9.10)$$

The objective function in Equation 9.3 minimizes the delivery cost by the reallocated IRs in hot spots with high demands to maximize the amount of offloaded traffic (that is, traffic delivered through the IRs). Equation 9.4 assures that the total number of IRs is less than or equal to the maximum number of IRs (i.e., N) that the network infrastructure can have. Equation 9.5 limits the allowed traffic to be offloaded to all IRs served by a publisher j to $1 - \beta_j$. The purpose of Equation 9.6 is to assure that each candidate region receives the requested demand. As indicated in the constraints, this can come from either the publisher or the reallocated IR.

Equation 9.7 defines the domain of the decision variables. Equation 9.8 assures the upper delay limit for all delivered contents. Finally, Equations 9.9 and 9.10 assure the feasibility of the found solution in terms of the IRs and original data publishers' capacities. By solving the formulated optimization problem in Equation 9.3, the optimal subset of candidate regions can be selected for the reallocation process, and the remaining hot spots can be served by the data publishers as summarized in Algorithm 9.2.

Algorithm 9.2: Optimized CDE-based approach

Function IRs-Reallocation (R, K, N)

Inputs

> R: Set of all candidate hot spots in the IoT,
> K: Set of all data publishers in the IoT,
> N: Total available IRs count.

Outputs

> *Cost: Total bandwidth consumed,*
> $\vec{l}$: Vector of load percentages to be dropped from data publishers' traffic load,
> $\vec{\alpha}$: Vector of selected hot spots for allocated IRs.

Begin

> Initialize: $\vec{P}, \vec{\gamma}, \vec{\beta}$ *and* μ.
>
> $[Cost, \vec{\alpha}$ *and* $\vec{l}]$ = Solve the ILP in Equations 9.3 through 9.10.
> Return
> $Cost, \vec{\alpha} = \{\alpha_1, \alpha_2, ..., \alpha_r\}$ *and* $\vec{l} = \{l_1, l_2, ..., l_r\}$.

End

9.5 Real Scenarios and Case Studies

In this section, four comprehensive and representative case studies for the different expected IoT traffic in reality have been considered as a proof of concept. Network demands and traffic behaviors, which line up with the published/used Internet traffic in Reference 34, have been observed and experienced at a university Web-service access level. These practical test cases represent a typical scenario for online servers that have been recently merged to be queried based on an information-centric system. This Web service supports more than 100,000 students and 2000 faculty. The main objective is to better and dynamically reallocate existing computing resources to meet the content demands of users/clients with the least cost. And thus, we model the observed content traffic and proactively allocate or redistribute the available network resources represented by IRs. Accordingly, all observed scenarios are classified into the following four cases.

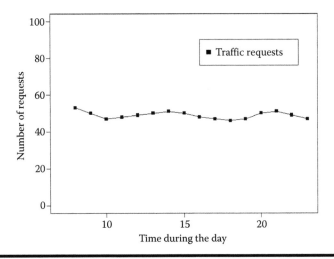

Figure 9.5   Steady network requests within the time unit.

Case 1: In an identified time period between 8 a.m. and 11 p.m., user requests arrive regularly and stay in a steady state, as demonstrated in Figure 9.5.

Case 2: In the same identified time period, users requests can peak consistently at certain times, as demonstrated in Figure 9.6, due to special events such as registration and/or add/drop periods.

Case 3: Another case can evolve due to the announcement of grades. Figure 9.7 illustrates typical network traffic for such a case. In this case, the announcement

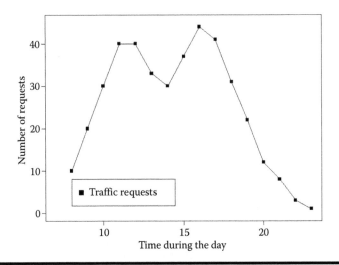

Figure 9.6   Fluctuating network requests within the time unit.

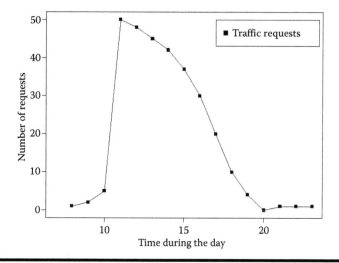

Figure 9.7 Some network requests within the time unit.

of grades was scheduled for 11 a.m., and, as expected, the network demand peaks at that time.

Case 4: Sometimes demand/service for content is low at the beginning but high toward the end of the time period. For instance, given a deadline for uploading the grades, the network activity increases dramatically. Figure 9.8 represents such a situation.

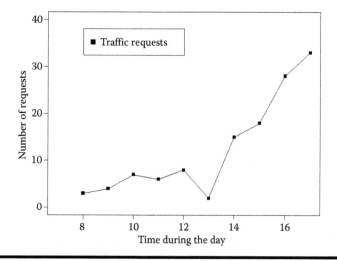

Figure 9.8 Drastic increase of network demand near deadline.

All the above mentioned cases, illustrated in Figures 9.5 through 9.8, have been used in the following section to assess the required IRs per case according to the CDE-based framework (see Figures 9.10a through 9.13a). Since the main characteristics of these cases' traffic can be precisely described based on the proposed CDE approach, we are able to observe and assess the allocated IRs in the examined information-centric IoT network (see Figures 9.10b through 9.13b).

9.6 Simulation Results and Discussions

In this section, Figures 9.9 through 9.20 display the simulation results of the proposed CDE-based approach based on the MATLAB (R2009b) programming environment. The *fit-ellipse*() function [35] is used to get the best-fit ellipse curve. For each case, we first plot the network requests, then extract the CDE characteristic curve, and finally, reverse the characteristic curves to advocate for the dynamic IR resource allocation amounts as shown in Figure 9.9, which depicts the changing requests in case 3, for instance.

The system is configured such that all the available resources are supplied at the beginning of the targeted time period in an attempt to make sure that the maximum demands are satisfied. Contrarily, and as shown in Figure 9.10, the proposed IR component allows flexible resource allocation that varies with demand to make the most gain out of limited resources in the IoT.

In line with Equation 9.1, we can conclude that the above-mentioned approach can be used to eliminate wastes in offered resources by simply adjusting the cost function such that all points outside the ellipse are weighted at a higher penalty.

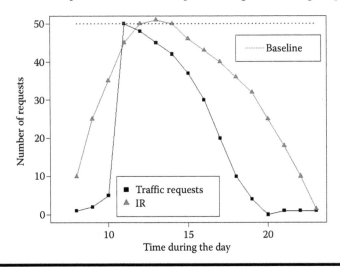

Figure 9.9 Overlay of requests, demand baseline, and dynamically allocated IRs for Case 3.

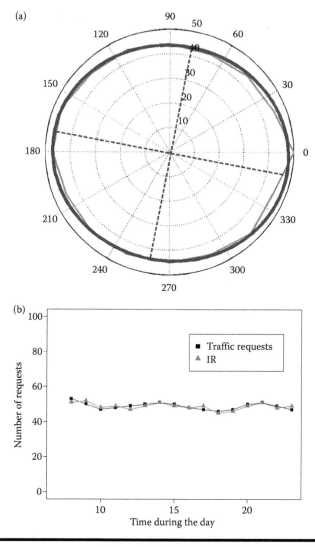

Figure 9.10 **(a) Mapping of network requests given in Figure 9.8 to polar coordinates and their best-fitted ellipse. (b) Resource requests and rectified curve of their best-fitted ellipse given in Figure 9.10.**

In order to show how the varying traffic behavior can affect the best-fitted ellipse parameters (such as the ellipse eccentricity [e], orientation angle [φ], and biased-center of the ellipse [d]), we examine the previously described case studies as follows.

Case 1: For this case, a steady state of traffic is assumed, and the best-fit ellipse is obtained as shown in Figure 9.10a. The ellipse parameters are also displayed in Table 9.2. Apparently, it has a low eccentricity value equal to 0.39; thus,

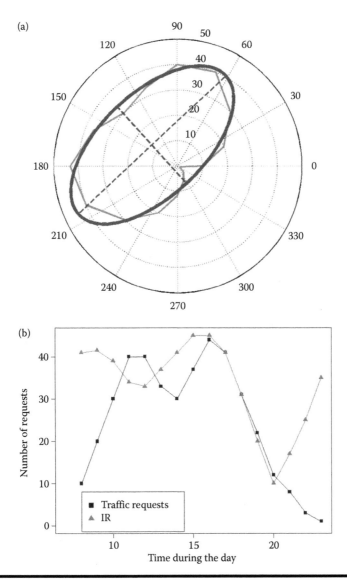

Figure 9.11 **(a) Mapping of network requests given in Figure 9.6 to polar coordinates and their best-fitted ellipse. (b) Resource requests and rectified curve of their best-fitted ellipse given in Figure 9.11a.**

the value of c parameter described above is very small, and the ellipse shape is almost a circle. Moreover, it is noticeable that the fitted-ellipse center is very close to the origin point with a very small d value (=0.49). The orientation angle of the ellipse is equal to $-10.72°$, which means that the direction of rotation of the ellipse is counterclockwise and there is almost a peak traffic demand at time $35T/36$. Consequently, this case concludes that when the network traffic is in a steady state, a pretty circular ellipse that is centered very close to the polar origin with a low d value is experienced. And once such a case is observed by the network operator, resources (i.e., IRs) can be allocated at the beginning of the period, and as time goes on, no reallocation will be required.

Case 2: A best-fit elliptical closed curve is obtained for this case, as shown in Figure 9.11a. As seen from the key parameters given in Table 9.4, the best-fitted ellipse has a high eccentricity of 0.8603. In addition, the centroid of the ellipse is away from the polar origin, with a high d value of 13.22. Furthermore, the orientation angle of the ellipse is equal to 42.4°. This indicates that a peak demand has occurred around the time $T/9$ and the rotation direction of the ellipse is clockwise. Note that in this example, T is 16 hours and demands begin to increase around 9:30 a.m.

Case 3: The best-fit ellipse of the closed curve for this case is shown in Figure 9.12a. As seen from the key parameters given in Table 9.4, the best-fitted ellipse has a high eccentricity of 0.7663. Its centroid is close to the polar origin, with a d value of 22.58. The orientation angle of the ellipse is equal to 43.55°, which means that the rotation direction is clockwise and there is a peak demand around time $T/8$.

Case 4: The best-fit ellipse's curve for this case is shown in Figure 9.13a. According to Table 9.4, the best-fitted ellipse has a high eccentricity of 0.8434. Although the eccentricity value is considered high, its centroid is relatively far from the polar origin, with a d value of 14.76. Meanwhile, the orientation of the ellipse is equal to −35.48°, which indicates a peak demand at $6T/7$ and a counter-clockwise ellipse rotation.

From the above-examined test cases, we can observe a positive orientation angle (or clockwise ellipse rotation) when the experienced IoT demands are increasing (cases 2 and 3) and a negative orientation when the demands are decreasing (cases 2 and 3). This can be a significant indicator at critical nodes such as those of the publishers and/or IRs for monitoring the network status per content traffic. It can be of utmost importance to learn the most appropriate caching locations in any information-centric paradigm. Furthermore, the high eccentricity and shift from

Table 9.4   Descriptive Parameters for the Best-Fitted Ellipses

	x_0	y_0	d	e	φ
Case 1	0.48	−0.5	0.49	0.3896	−10.72°
Case 2	−10.31	8.2845	13.2266	0.8603	42.4°
Case 3	−6.70	21.56	22.58	0.7663	43.55°
Case 4	7.6862	−12.6047	14.76	0.8434	−35.48°

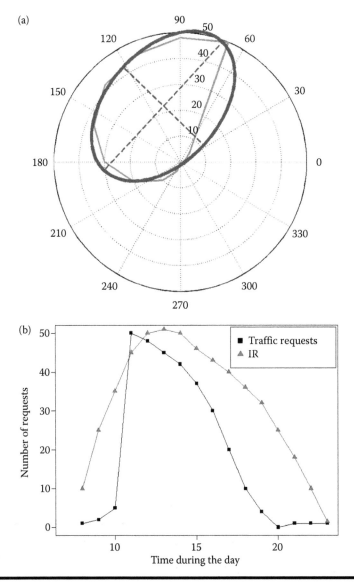

Figure 9.12 (a) Mapping of network requests given in Figure 9.7 to polar coordinates and their best-fitted ellipse. (b) Resource requests and rectified curve of their best-fitted ellipse given in Figure 9.12a.

the polar origin indicate an imbalance in the content demand; thus, cases 2, 3, and 4 necessitate a dynamic reallocation of the IRs for more optimized cost and efficient traffic handling. For example, cases 3 and 4 have the highest center bias d equal to 22.58 and 14.76, respectively. Consequently, additional publishers in the form of IRs must be reallocated. It is worth pointing out also that the higher the

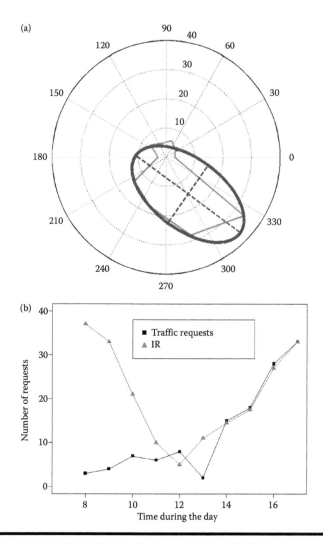

Figure 9.13 **(a) Mapping of network requests given in Figure 9.5 to polar coordinates and their best-fitted ellipse. (b) Resource requests and rectified curve of their best-fitted ellipse given in Figure 9.13a.**

eccentricity value is, the more attention has to be given to balance the demands in information-centric networks.

Accordingly, the percentage of savings is calculated for the previously discussed cases, as shown in Table 9.5. The smallest percentage of savings (8.5%) is obtained from case 1 when the network traffic is stable and not too much optimization can be applied. Nevertheless, when the demand fluctuates drastically as shown in cases 2, 3, and 4, the eccentricity increases; thus, the potential for more optimization and savings (~35%) can be achieved.

Table 9.5   Total Requests, Baseline, and IRs Count

	Request	*Baseline*	*IRs*	*Allocation Savings (%)*
Case 1	725	800	731	8.5680
Case 2	400	800	527	34.0464
Case 3	297	800	536	32.9362
Case 4	122	340	208	38.6638

Therefore, in the proposed ILP described above, a limited number of IRs is proposed for cost effectiveness, in which the ILP responds to the incoming requests analyzed in the aforementioned four cases by virtually reallocating the IRs in the network. By running a CPLEX-based MATLAB simulator, the number of available repeaters N versus the *Cost* curve is obtained and analyzed in Figure 9.14. Obviously, adding more IRs increases the setup cost, which is a one-time fee. Nevertheless, the objective here is to confirm that increasing the IRs in information-centric networks can reduce the network cost in terms of bandwidth when multiple data publishers/repeaters are utilized. Furthermore, with lower values of r, we can reach zero cost faster. On the other hand, more IRs should be added to get the same effect with lower values of r. Also, we can see that cost decreases to 0 when the total IR and r counts reach the equilibrium state.

In Figure 9.15, we give higher load weights to the first two data publishers than the others, while distributing the data requests' loads randomly, even though this didn't affect the curve pattern, which makes us more confident about the behavior

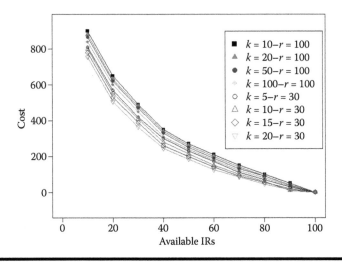

Figure 9.14   Total IRs count vs. the cost function.

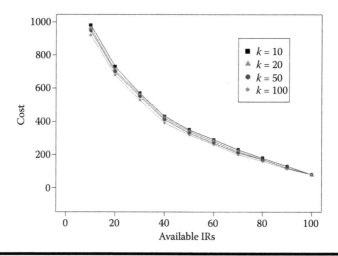

Figure 9.15 Total IRs vs. the cost with weighted randomly generated P_is.

of the proposed CDE-based approach. It confirms that this approach is so efficient that even though we have unbalanced publishers, our cost is still converging to zero.

From Figure 9.16, we can observe that the cost is still reasonable while activating the IRs using the proposed CDE-based approach. While assuming r is equal to 100 regions in Figure 9.16 and k is increasing from 10 to 100, we can conclude that adding IRs can save up to 92% of the cost. This is confirmed from the upper curve in Figure 9.16, which accumulates traffic while no IRs are considered. Moreover, when IRs are placed/activated, we can observe that the cost is slowly decreasing to zero.

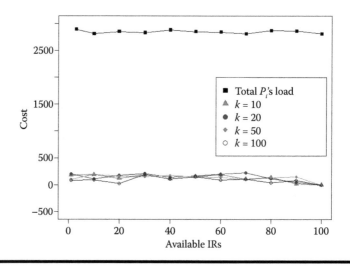

Figure 9.16 Total IRs vs. the cost in comparison to summation of P_is.

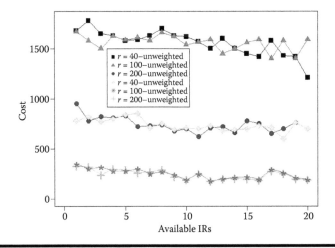

Figure 9.17 Total IRs vs. the cost function with balanced and unbalanced publishers' load.

In Figure 9.17, we examine the unbalanced versus balanced publisher loads. We vary the hot spots' counts in order to see the difference in cost. The CDE-based approach again shows its efficiency even with unbalanced publishers' loads, which is often the case in practice. Consequently, at the end, while adding the 20th IR, the cost of the weighted/unweighted publishers' loads doesn't vary much while assuming a lower count of hot spots. However, there is a significant cost change between the unbalanced and balanced publisher loads when the r value is equal to 200.

In Figure 9.18, we examine the percentage of total network traffic off-loaded with respect to the count of deployed IRs. The assumed IoT network with 30 and 100

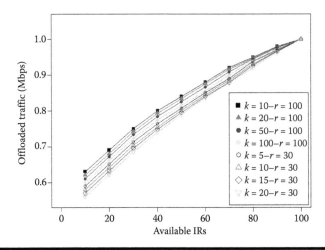

Figure 9.18 Total IRs count vs. the off-loaded traffic.

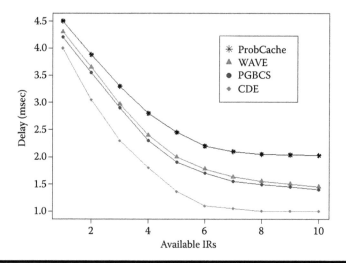

Figure 9.19   Experienced average delay as a function of the IRs count.

hot-spot cases in which k values vary from 5 to 20 publishers is inspected. Obviously, with lower numbers of hot spots, we can achieve the 100% off-loaded traffic much faster. For example, with 30 hot spots, 100% off-loaded traffic has been achieved much faster than it would have been with 100 hot spots. We can conclude that the lower the hot-spot number is, the faster it off-loads the network traffic. However, it necessitates more IRs when we increase the hot spots' demanding regions.

In Figure 9.19, average end-to-end is examined with respect to the total used IRs' count. It is clear that with a few more IRs, the experienced delay can be decreased

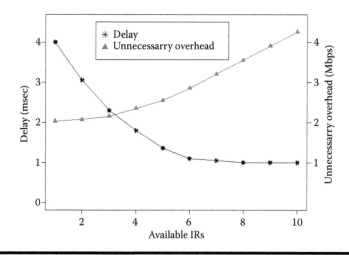

Figure 9.20   Delay vs. the unnecessary overhead traffic as a function of the IRs' count.

dramatically. Furthermore, we can observe that there is a steady decrease in delay after adding the 6th and 7th IRs. After adding the 8th IR, the delay reaches the value of one msec approximately. Reaching this competitive delay with a lower number of IRs proves that the proposed CDE-based framework is significantly cost-efficient. Moreover, in the same figure (Figure 9.19), the CDE approach is compared to other related approaches in the literature, which rely on caching in providing closer replicas to the end user and include approaches such as ProbCache [36], Cost-aware [37], and Wireless Access Virtual Environment (WAVE) [38]. ProbCache is a decentralized caching approach for real-time content distribution in routers' caches. It approximates the caching capability of a path and caches contents probabilistically in order to better utilize the available caching space on the same path and fairly multiplex contents of different flows in caches along a shared path. In Reference 37, the authors proposed a cost-aware approach that takes into consideration the cached content's popularity and the cost gain in caching. This approach is called popularity and gain-based caching scheme (PGBCS), and it realizes the placement and replacement of cached contents through comprehensive measurements. This approach can effectively promote the node's cache hit ratio and reduce the user request delay; thus, it has been chosen in this study to be one of the baselines for comparison. Meanwhile, WAVE adjusts replicas to be cached based on their popularity. An upstream node recommends the number of replicas to be cached at its downstream node, which is exponentially increased as the request count increases. Since the average hop count for the content delivery using WAVE is claimed to be very low, we are using it as a competitive baseline to the proposed delay-tolerant CDE approach. It is obvious from Figure 9.19 how competitive the aforementioned baselines are in terms of delay to the proposed CDE-based approach. However, the CDE approach still outperforms the other three approaches due to the accurate elliptical estimation of the content traffic per IR in the network.

In Figure 9.20, we elaborate more on the efficiency of the CDE approach in making more optimized decisions about the experienced delay factor. By plotting the average experienced delay against the unnecessary overhead that is experienced due to IRs' management/synchronization traffic, we can observe a significant trade-off between these two information-centric design factors. We can decide on an optimum configuration for the scenario in Figure 9.20, in which the recommended IRs count should be equal to three IRs in order to maintain the lowest overhead for the best gain in terms of delay. Accordingly, this figure shows how the proposed CDE approach lends itself as an important tool in effectively specifying the operative IRs' count range, considering the unnecessary overhead traffic as well as the QoS in terms of delay.

9.7 Conclusion

In this chapter, a CDE framework is introduced for IoT networks. CDE is based on a novel architecture that proposes a new entity called IRs in addition to typical IoT entities such as the data publishers and/or intermediate nodes. According to the proposed

CDE framework, IRs are redistributed based on the most up-to-date consumers' requests. CDE implements two algorithms that utilize elliptical properties in representing the irregularity of content demands. It provides a dynamic IoT architecture that acts based on both ends of the publisher-consumer chain. A detailed performance evaluation of the proposed reallocation strategy was conducted, and the trade-offs between delivery cost and IRs' utilization were discussed. Moreover, real case studies of varying content request traffic have been examined to show how the key parameters of the CDE can be used to model the network behavior. Please note that the key parameters of CDEs can be used to make intelligent decisions on resource reallocation (e.g., the IRs). Furthermore, the proposed ellipse-based model uses fewer parameters than other traditional methods in modeling the IoT traffic. It is worth pointing out that by using properties of the ellipse, the network can be trained to adaptively construct an IR placement algorithm with artificial intelligence (AI) techniques.

References

1. L. Atzori, A. Iera, and G. Morabito, The internet of things: A survey, *Computer Networks*, vol. 54, no. 15, pp. 2787–2805, 2010.
2. J. Wu and I. Stojmenovic, Ad hoc networks, *Computer-IEEE Computer Society*, vol. 37, no. 2, pp. 29–31, 2004.
3. F. Al-Turjman, Information-centric sensor networks for cognitive IoT: An overview, *Annals of Telecommunications*, vol. 72, no. 1, pp. 3–18, 2017.
4. A. Ghodsi et al., Information-centric networking: Seeing the forest for the trees, in *Proceedings of the 10th ACM Workshop on Hot Topics in Networks*, ACM, New York, NY, 2011.
5. B. Ahlgren et al., A survey of information-centric networking, *IEEE Communications Magazine*, vol. 50, no. 7, pp. 26–36, 2012.
6. Internet Research Task Force, Information-Centric Networking Research Group, [online]: https://irtf.org/icnrg
7. F. Al-Turjman, Price-based data delivery framework for dynamic and pervasive IoT, *Elsevier Pervasive and Mobile Computing Journal*, 2017. doi: 10.1016/j.pmcj.2017.05.001
8. G. Xylomenos et al., A survey of information-centric networking research, *IEEE Communications Surveys & Tutorials*, vol. 16, no. 2, pp. 1024–1049, 2014.
9. Cisco's annual report [online]: http://www.cisco.com/c/en/us/about/annual-reports.html
10. J. Sahoo, S. Cherkaoui, and A. Hafid, Optimal selection of aggregation locations for participatory sensing by mobile cyber-physical systems, *Computer Communications*, vol. 74, pp. 26–37, 2016.
11. W. Ibrahim, A. Taha, and H. Hassanein, Using smart vehicles for localizing isolated Things, *Computer Communications*, vol. 74, pp. 16–25, 2016.
12. A.M. Mohammed and A.F. Agamy., A survey on the common network traffic sources models, *International Journal of Computer Networks*, vol. 3, no. 2, 2011.
13. A.J. Field, U. Harder, and P.G. Harrison, Measurement and modelling of self-similar traffic in computer networks, *IEE Proceedings Communications*, vol. 151, no. 4, pp. 355–363, 2004.

14. V. Paxson and S. Floyd, Wide area traffic: The failure of Poisson modeling, *IEEE/ACM Transactions on Networking*, vol. 3, no. 3, pp. 226–244, 1995.
15. G. Singh and F. Al-Turjman, A data delivery framework for cognitive information-centric sensor networks in smart outdoor monitoring, *Elsevier Computer Communications*, vol. 74, pp. 38–51, 2016.
16. F. Al-Turjman, Cognitive-node architecture and a deployment strategy for the future sensor networks, *Springer Mobile Networks and Applications*, 2017. doi: 10.1007/s11036-017-0891-0
17. INFSO D.4 Networked Enterprise & RFID INFSO G.2 Micro & Nanosys., in co-operation with the Working Group RFID of the ETP EPOSS, *Internet of Things in 2020, Roadmap for the Future*, Ver. 1.1, May 2008.
18. F. Al-Turjman, Cognitive caching for the future fog networking, *Elsevier Pervasive and Mobile Computing*, 2017. doi: 10.1016/j.pmcj.2017.06.004
19. A. Dan and D. Towsley, *An Approximate Analysis of the LRU and FIFO Buffer Replacement Schemes*, ACM, vol. 18, no. 1, 1990.
20. X. Vasilakos, V.A. Siris, and G.C. Polyzos, Addressing niche demand based on joint mobility prediction and content popularity caching, *Computer Networks*, vol. 1, no. 10, pp. 306–323, 2016.
21. F. Zhang, C. Xu, Y. Zhang, and K. Ramakrishnan, EdgeBuffer: Caching and prefetching content at the edge in the MobilityFirst future Internet architecture, in *Proceedings of the IEEE International Symposium on World of Wireless, Mobile and Multimedia Networks*, Boston, MA, pp. 1–9, 2015.
22. E. Bastug, M. Bennis, and M. Debbah, Living on the edge: The role of proactive caching in 5G wireless networks, *IEEE Communications Magazine*, vol. 52, no. 8, pp. 82–89, 2014.
23. N. Golrezaei, K. Shanmugam, A.G. Dimakis, A.F. Molisch, and G. Caire, Femtocaching: Wireless video content delivery through distributed caching helpers, in *Proceedings of the IEEE International Conference on Computer Communications*, 2012.
24. P. Bahl, R. Chandra, P.P.C. Lee, V. Misra, J. Padhye, D. Rubenstein, and Y. Yu, Opportunistic use of client repeaters to improve performance of WLANs, *IEEE/ACM Transactions on Networking*, vol. 17, no. 4, pp. 1160–1171, 2009.
25. J. Xiong and R.R. Choudhury, Peercast: Improving link layer multicast through cooperative relaying, in *Proceedings of the IEEE International Conference on Computer Communications*, pp. 2939–2947, April 2011.
26. P. Frangoudis, G. Polyzos, and G. Rubino, Content dissemination in wireless networks exploiting relaying and information-centric architectures, in *Proceedings of the International Workshop on Quality, Reliability, and Security in Information-Centric Networking (Q-ICN)*, Rhodes, Greece, August 2014.
27. F. Al-Turjman, H. Hassanein, and M. Ibnkahla, Optimized relay repositioning for wireless sensor networks applied in environmental applications, in *Proceedings of the IEEE International Wireless Communications and Mobile Computing Conference (IWCMC11)*, Istanbul, Turkey, pp. 1860–1864, Jul. 2011.
28. F. Al-Turjman, H. Hassanein, and M. Ibnkahla, Efficient deployment of wireless sensor networks targeting environment monitoring applications, *Elsevier: Computer Communications Journal*, vol. 36, no. 2, pp. 135–148, 2013.
29. F. Al-Turjman, Optimized hexagon-based deployment for large-scale ubiquitous sensor networks, *Springer's Journal of Network and Systems Management*, 2017. doi: 10.1007/s10922-017-9415-2

30. D. Turgut and L. Boloni, Heuristic approaches for transmission scheduling in sensor networks with multiple mobile sinks, *The Computer Journal*, vol. 54, no. 3, pp. 332–344, 2011.

31. V. Sourlas, L. Tassiulas, I. Psaras, and G. Pavlou, Information resilience through user-assisted caching in disruptive content-centric networks, in *Proceeding of IFIP Networking*, pp. 1–9, 2015.

32. F. Al-Turjman and H. Hassanein, Enhanced data delivery framework for dynamic Information-Centric Networks (ICNs), in *Proceedings of the IEEE Local Computer Networks (LCN)*, Sydney, Australia, pp. 831–838, 2013.

33. D. Alderson, H. Chang, M. Roughan, S. Uhlig, and W. Willinger, The many facets of internet topology and traffic, *Networks and Heterogeneous Media*, vol. 1, no. 4, pp. 569–600, 2006.

34. Internet traffic dataset [online]: http://www.maths.adelaide.edu.au/matthew.roughan/project/traffic_matrix/

35. O. Gal, MATLAB central – File detail – fit_ellipse. MATLAB Central, 2003. Available at: http://www.mathworks.com/matlabcentral/fileexchange/3215-fit-ellipse (accessed April 9, 2015).

36. I. Psaras, W.K. Chai, and G. Pavlou, Probabilistic in-network caching for information-centric networks, in *Proc. of the ICN workshop on Information-centric networking*, Helsinki, Finland, pp. 55–60, 2012.

37. Z. Fan, Q. Wu, M. Zhang, and R. Zheng, Popularity and gain based caching scheme for information-centric networks, *International Journal of Advanced Computer Research*, vol. 7, no. 30, pp. 71–80, 2017.

38. K. Cho, M. Lee, K. Park, T.T. Kwon, and Y. Choi, WAVE: Popularity-based and collaborative in-network caching for content-oriented networks, in *Proceedings of the IEEE INFOCOM Workshop on Emerging Design Choices in Name-Oriented Networking*, Orlando, FL, USA, pp. 316–321, 2012.

Chapter 10

Conclusions and Future Directions

Multimedia and mobile applications have emerged as one of the most promising wireless sensor network (WSN)-enabled technologies in the Internet of things (IoT) era. The IoT domain has been spread across a variety of areas, including the future Internet, object identification using radio-frequency identification (RFID), and various sensing techniques to improve the network performance and quality of experience. In this work, we proposed and evaluated the use of multimedia-enabled sensor networks in large-scale IoT applications. We focused on key design aspects in data delivery and network traffic.

10.1 Summary of the Book

We started the work in this book with a comprehensive overview about the use of multipath routing in wireless multimedia sensor networks (WMSNs) while discussing its applicability under the IoT circumstances in Chapter 2. The vision of WMSNs is to provide real-time multimedia applications using wireless sensors deployed for long-term usage. However, quality of service (QoS) assurances for both best-effort data and real-time multimedia applications introduced new challenges in prioritizing multipath routing protocols in WMSNs. Multipath routing approaches with multiple constraints have received considerable research interest. In Chapter 2, a comprehensive survey of both best-effort data and real-time multipath routing protocols for WMSNs is presented. The results of a preliminary investigation into design issues affecting the development of strategic multipath routing protocols that support multimedia data in WMSNs are also presented and discussed from the network application perspective. From the obtained results, we concluded with the capability of multipath routing in the IoT era.

In Chapter 3, we propose a mathematical model for a novel QoS routing determination method. The proposed scheme determines the optimal path that provides an appropriate shared radio satisfying the QoS requirements for a wide range of real-time intensive media. The mathematical model is based on the Lagrangian relaxation method, to control adaptive switching of hop-by-hop QoS routing protocols. The embedded criteria for each objective function are used to decide which path from source to sink will be selected. Simulation results show that, compared with existing routing protocols, the approach proposed in this chapter significantly improves the packet received ratio, energy consumption, and average end-to-end delay of the sensor node.

In Chapter 4, we consider the development of a hybrid sensor and vehicular network (HSVN) that provides safety supports to intelligent transportation systems (ITS) and assists emergency situations in the event of a disaster via multimedia exchange. Our main interest is emergency situations in a smart city and its facilities during and/or after disasters such as earthquakes and severe car collisions, where unsafe road segments are often found. We propose an agile framework that caters to service-based applications in smart cities where multimedia data are heavily exchanged. An optimized data delivery approach that operates with limited resources in highly dynamic topologies is investigated and proposed. We also propose a solid mathematical model for a novel QoS routing-determination method. This model helps in specifying which path to follow in order to determine the optimal usage of the available resources while satisfying QoS for a wide range of real-time multimedia applications in safety and security. Simulation results, which have been validated via solid analytical analysis, are used to assess the efficiency of the proposed approach in terms of the packet received ratio, energy consumption, and average end-to-end delay.

In Chapter 5, we introduce a framework for node deployment and delay tolerance in RFID-sensor networks (RSNs) under the IoT paradigm. Our framework is comprised of two components. The first is a novel smart-integrated WSN and RFID (SIWR) architecture that classifies nodes into light nodes and super integrated nodes. SIWR employs an optimal linear programming formulation to cost-efficiently place integrated RFID readers. However, integrated IoT architectures face significant connectivity challenges. The IoT settings assume no guaranteed contemporaneous end-to-end path between node pairs. To this end, in the second part of our framework, we introduce an optimized delay-tolerant approach for integrated RFID-sensor networks (DIRSNs). This is a novel scheme for data routing and courier nodes' selection in RSNs. The DIRSN's formulation minimizes delay across the network without violating the main dense-deployment and load-balancing requirements. In addition, DIRSN builds on SIWR's novel architecture to locate the best set of couriers that promise to provide connectivity. Our combined approach is compared to three types of RSN integration architectures, and the results show that our architecture and courier selection approaches perform substantially better than other architectures in terms of minimizing delay, cost, and packet loss and in handling extensive traffic demands.

In Chapter 6, our main interest is the emergency situations in a smart city and its facilities during and/or after disasters such as earthquakes and severe car collisions, where unsafe road segments are often found. For recovering, constructing, and selecting *k*-disjoint paths capable of putting up with failure of the parameters but satisfying the QoS, we propose a bio-inspired particle multipath swarm optimization (PMSO) multimedia delivery framework. During communication with the unmanned aerial vehicle (UAV), the multi-swarm strategy is used to determine the optimal direction while performing a multipath routing. Authenticity of the proposed approach has been tested and results show that, compared to the canonical particle swarm optimization (CPSO) and fully particle multipath swarm optimization (FPMSO), the proposed method is better.

In Chapter 7, asynchronous medium access control (MAC) duty-cycled protocols have been assessed and evaluated as superior due to their high energy efficiency and reduced packet latency in comparison to synchronized ones. This type of protocol is considered very important in MAC protocols due to the adverse effects of hidden terminals that cause energy consumption in sensor networks. Therefore, in this chapter, the impact of hidden terminals on the performance of an asynchronous duty-cycled MAC protocol X-MAC for vehicle-based sensor is investigated via analysis and simulations. We propose a Markov model to analyze the QoS parameters in terms of energy consumption, delay, and throughput. Our analytical model provides QoS parameter values that closely match the simulation results under various network conditions. Our model is more computationally efficient and provides accurate results quickly compared with simulations. More importantly, our model enables the designers to obtain a better understanding of the effects of different numbers of mobile sensor nodes and data arrival rates on the performance of an asynchronous MAC duty-cycled protocol.

In Chapter 8, we present a novel traffic model for a new generation of sensor networks that supports a wide range of communication-intensive real-time multimedia applications. The model is used to investigate the effects of multi-hop communication on ITSs via a Markov discrete-time M/M/1 queuing system. Moreover, an analytical formulation for the bit error rate (BER) and the critical path loss model is presented. We address the degree of irregularity for location-based switching with respect to two categories in distributed retransmissions: the hop-by-hop and the end-to-end retransmissions. Simulation results based on a realistic case study and assumptions are performed to highlight the effects on the average packet delay, energy consumption, and network throughput. The findings presented in this work are of great help to designers of WMSNs.

With the increased growth in the number of connected devices, either static or mobile ones, there is a concurrent massive increase in the accompanied data traffic volume. Therefore, and for better future communication systems with better coverage and capacity performance, the information-centric IoT is a prudent option. In this IoT paradigm, populating and reallocating information repeaters (IRs) is one promising way to reduce data traffic during the peak periods. Accordingly, a novel placement approach for the IRs in high-demand regions (hot spots) is introduced in Chapter 9. Placement of the IRs in these regions is prioritized while sustaining three main objectives which aim at (i) reducing network traffic, (ii) maintaining an upper bound for the experienced delay, and (iii) minimizing the average traffic load per publisher. Moreover, this research examines the performability of the IoT paradigm under varying traffic behaviors by using a new method in characterizing the network behavior based on content demand ellipses (CDEs). Real data traffic has been used to show how various traffic behaviors can affect key properties/parameters in the proposed CDE representation, and promising results have been achieved.

10.1.1 Future Directions

With the research done in this book, WMSNs will be able to provide better infrastructure support for smarter paradigms across the globe. In addition, several future research directions and open issues can also be derived from the work done in this book so far. In the following, we outline some of these future directions and open research issues.

1. *Exploiting the relay placement problem more and more:* One of the key aspects in the WMSNs is the positioning of the multimedia relays and their availability across the network. Such relays would raise an issue with the mobility-enabled node presence. Thus, exploring the role of mobile relays in multimedia delivery and the study of their replacement techniques that suit the multimedia sensor nodes in the mobile IoT is still a direction to explore out of this work.

2. *Mobility-enabled sensor nodes:* In general, the effect of node mobility on the QoS and its impact on the WMSN adaptability and longevity should be investigated more for safer IoT applications, such as those found in ITSs and health care. While the discussed mobile WMSN architecture can help in overcoming the limitations of the cross-layer design, we strongly recommend further investigations in the physical layer components and configurations.

3. *More developed and commercialized cognitive multimedia relays:* In terms of enhancing the multimedia delivery while supporting diverse application platforms, this work can be reconsidered for further enhancement in domain-specific technology for better multimedia relaying in the IoT era. The creation of such technology can contribute toward the development of an enterprise architecture that can be applied to different application domains by using the same underlying WMSN.

4. *WMSN integration with the next-generation networks (NGNs):* More functions could be incorporated at the level of the sensor node to integrate it with the next generation networks (NGNs), such as the 5G network, moving toward more multimedia-enabled nodes, working in smart network setups. The expansion of the relay functions would then be able to take requests directly from different network users, such as cell phones, access points, and base stations, thus making the WMSN platform more accessible to end users in the IoT paradigms.

5. *Cognition in the WMSN:* The idea of cognition can be investigated while being applied to intermediate sensor nodes of the current WMSN infrastructure to realize the cognitive network concept in general. Multimedia need not be requested from specific hosts nor has to travel end-to-end in the network. Instead, it can be cached at the network's edge (or Fog). Cognitive relays could be used to understand the user request patterns and manage the cache content intelligently.

6. *Security and privacy open issues:* Security and privacy are key issues to be addressed while spreading the adoption of WMSN in IoT applications. Unluckily, security solutions have not been incorporated in a planned way in the early IoT version. IoT-enabling technologies such as RFID and sensor networks were simply integrated across the existing Internet infrastructure without considering key design factors such as trust and privacy issues while gathering and delivering multimedia traffic. These security considerations span not only the multimedia management, application, and service levels but also the communication and networking levels; these are significant aspects to be investigated in any future work.

Index